The Crisis Management Cycle

The Crisis Management Cycle is the first holistic, multidisciplinary introduction to the dynamic field of crisis management theory. By drawing together the different theories and concepts of crisis management literature and practice, this book develops a theoretical framework of analysis that can be used by both students and practitioners alike. Each stage of the crisis cycle is explored in turn:

- Risk assessment
- Prevention
- Preparedness
- Response
- Recovery
- Learning

Stretching across disciplines as diverse as safety studies, business studies, security studies, political science and behavioural science, *The Crisis Management Cycle* provides a robust grounding in crisis management that will be invaluable to both students and practitioners worldwide.

Christer Pursiainen is Professor of Societal Safety at the Arctic University of Norway (UiT), Tromsø, Norway. He has published widely on a variety of themes such as crisis decision-making, critical infrastructure protection and resilience, international relations theory, foreign policy analysis, regional cooperation and integration, and comparative politics.

The Crisis Management Cycle

Christer Pursiainen

Routledge
Taylor & Francis Group
LONDON AND NEW YORK

First published 2018
by Routledge
2 Park Square, Milton Park, Abingdon, Oxon OX14 4RN

and by Routledge
711 Third Avenue, New York, NY 10017

Routledge is an imprint of the Taylor & Francis Group, an informa business

© 2018 Christer Pursiainen

The right of Christer Pursiainen to be identified as author of this work has been asserted by him in accordance with sections 77 and 78 of the Copyright, Designs and Patents Act 1988.

All rights reserved. No part of this book may be reprinted or reproduced or utilised in any form or by any electronic, mechanical, or other means, now known or hereafter invented, including photocopying and recording, or in any information storage or retrieval system, without permission in writing from the publishers.

Trademark notice: Product or corporate names may be trademarks or registered trademarks, and are used only for identification and explanation without intent to infringe.

British Library Cataloguing in Publication Data
A catalogue record for this book is available from the British Library

Library of Congress Cataloging in Publication Data
Names: Pursiainen, Christer, author.
Title: The crisis management cycle / Christer Pursiainen.
Description: Abingdon, Oxon ; New York, NY : Routledge, 2018. | Includes bibliographical references and index.
Identifiers: LCCN 2017031172 | ISBN 9781138643871 (hardback) | ISBN 9781138643888 (pbk.) | ISBN 9781315629179 (ebook)
Subjects: LCSH: Crisis management. | Emergency management. | Conflict management. | Risk management.
Classification: LCC HD49 .P87 2018 | DDC 658.4/056--dc23
LC record available at https://lccn.loc.gov/2017031172

ISBN: 978-1-138-64387-1 (hbk)
ISBN: 978-1-138-64388-8 (pbk)
ISBN: 978-1-315-62917-9 (ebk)

Typeset in Times New Roman
by Taylor & Francis Books

Contents

List of illustrations   vi
Preface   vii

1	Introduction	1
2	Risk assessment	9
3	Prevention	39
4	Preparedness	69
5	Response	96
6	Recovery	127
7	Learning	146
8	Conclusions	169
	Index	179

Illustrations

Figures

1.1	The crisis management cycle	5
2.1	ISO 31000 risk management framework	13
2.2	Bow-tie diagram	27
2.3	Fault Tree Analysis	28
2.4	Event Tree Analysis	28
3.1	Bow-tie diagram with barriers	50
3.2	The ALARP principle	60
4.1	Passive and active redundancy	83
6.1	Resilience of three systems	133
7.1	Levels of post-crisis learning	155

Tables

2.1	Risk scenario with varying conditions	33
2.2	Risk matrix	35
3.1	Early warning vulnerability factors	51
5.1	Decision-making theories	105
7.1	Factors contributing to learning failures	156
8.1	Crisis management discourses and debates	175

Preface

When I took over the Professorship in Societal Safety and Environment at the Arctic University of Norway in 2014, I inherited a course called Crisis Management. This was rather convenient as I had dealt with crisis management throughout my professional career in different positions and roles, and was therefore familiar with most of the issues that are central to this field. When preparing the course, I noticed, however, that the field is very fragmented, as was my knowledge at the time.

The usual way to teach this field is to present the students with some key texts, which are often unrelated to each other. I decided I needed a more holistic framework to make sense of the diverse approaches and their relationships to each other. Before long, it dawned on me that the crisis management cycle – not at all a new invention, although not so often applied in theoretical contexts – provides this kind of framework, within which basically all the crisis management-related literature can be situated.

Teaching, that is, speaking aloud, is a good way to clarify things for oneself that one thinks one knows, and to find gaps in one's knowledge. One quickly realises that giving a three-hour lecture on, say, crisis prevention or crisis communication, requires considerable material and preparation, and a good overview and clear perspective on the theme. This led me to collect and classify the existing literature more systematically according to the crisis management cycle phases. During this process, I benefited greatly from interaction with the students. After running the course twice, I thought that a book that would support my course in terms of providing a holistic and integrated view of crisis management would not be out of place.

Hence the current book. I hope it will benefit not only my students, but the widest possible audience, including newcomers to the field and experienced practitioners alike who have to deal with some aspects of crisis management, irrespective of their academic discipline or professional branch.

I would like to thank all of my colleagues, both at my university and in different projects I am involved in, for the insights they have contributed to different parts of the current book. In addition, I am grateful for having been able to cooperate closely over the years with crisis management experts, especially in the fields of civil protection, nuclear and radiation safety, and law enforcement, from local to national and international levels. I am also indebted to the EU Horizon 2020 IMPROVER Project (grant agreement no. 653390), within which I have been able to apply the crisis management cycle approach to the related concept of resilience. Without this experience and the knowledge I have gained, the book would not be the same as it is now. Finally, I am grateful to Angelica Matveeva, who helped me to streamline parts of the text, and to Lynn Nikkanen, who edited the language in the first version of the manuscript.

Christer Pursiainen,
Tromsø, June 2017

1 Introduction

Crisis management is a broad field. It is relevant to many contexts and disciplines, such as societal security, political science, public policy and administration, international relations, business studies, organisation studies, military studies, psychology, behavioural science, environmental science, health sciences, engineering, and so forth. It is therefore rather striking that while there exists an abundance of literature on individual elements of crisis management, there is a relative lack of more holistic treatments of this theme.

Most of the relatively few book-length treatments of crisis management are limited to one particular field or discipline. The overwhelming majority of these books focus on two areas. Most of them deal with civil emergencies from the perspective of public administration and political leadership (e.g. Drennan, McConnell and Stark, 2015; McEntire, 2015; Deverell, Hansen and Olsson, 2015; Farazamand, 2014; Watters, 2014; Fagel, 2013; Haddow, Bullock and Coppola, 2011; Boin et al., 2005; Rosenthal, Charles and 't Hart, 1998). There are also a few holistic books on corporate reputation management issues (e.g. Crandall, Parnell and Spillan, 2014; Dezenhall and Weber, 2011; Regester and Larkin, 2008; Harvard Business School, 2004; Fink, 2002). This type of crisis literature is written for a variety of purposes and audiences, and can roughly be divided into practical manuals, edited compilations of case studies or collections of mainly pre-existing journal articles.

By contrast, there is considerable conceptual discussion around the theme in the form of academic articles. Since the 1990s in particular, this literature has been growing at a steady pace and there have been calls for the development of 'new approaches in crisis management' in order to cope with the changing world and conditions (e.g. Quarantelli, 1996; Robert and Lajtha, 2002; Boin and Lagadec, 2000).

Is it possible to create some kind of synthesis of this development, which seems to be characterised by efforts to become a field in its own right, but at the same time exhibits all the hallmarks of becoming increasingly fragmented? The current book is written in the belief that there is a future for crisis management as an established academic and practical field. To this end, the book aspires to provide a generic account of crisis management applicable to any country and any discipline, any society and any organisation, combining old and new approaches, theory and practice. For students and their teachers, it offers a theoretical and methodological framework for analysis, being applicable to all disciplines where crisis management is taught and studied. For practitioners, it offers a framework for action that can be tailored to all contexts where crises occur, and where their successful management is desirable and possible.

1.1 What is crisis management?

The concept of crisis is used in a variety of ways depending on the discipline, field or context. While some convergence of the definition has emerged in theoretical and applied literature, it has been noted that the lack of a common definition of what constitutes a crisis undermines the development of crisis management as a theoretical field (Roux-Dufort and Lalonde, 2013). Thus, from a corporate point of view, a crisis can be seen as a turning point that can lead a company in either a better or worse direction (Fink, 2002). From the public management point of view, it can be seen as a situation that is beyond the capacity of normal management structures and processes to deal with effectively (e.g. the Australian definition, in CIPedia, n.d.). From an organisational perspective, it has been argued that, in many cases, a situation is defined as a crisis when, in actual fact, it is only a question of a controversy among stakeholders; unlike a crisis, which is a major urgent event that impacts reputation, controversy emerges from the way in which communication within a network of organisations shapes an issue (Ziek, 2015).

The picture is further complicated by the fact that there are several other terms that overlap, and that are sometimes equated with the concept of crisis. These include terms such as catastrophe, disaster, emergency, incident, event, and so forth. Unlike the concept of crisis, most of these terms are rather standardised (e.g. UNISDR, 2009; ISO 2009a, 2009b; ISO/PAS, 2007; UNDHA, 1992, p. 34). Some civil emergency management researchers are furthermore inclined to position a crisis somewhere between accidents classified as minor in terms of their event magnitude, on the one hand, and emergencies, disasters, calamities and catastrophes, on the other, the latter being much more severe in their magnitude than crises (McEntire, 2015, p. 3).

Nevertheless, from the perspective of a multidisciplinary discussion of crisis management, the broadest and most inclusive definition of the concept of crisis proves to be the most useful. When one looks at the different definitions, there are at least three key criteria, as identified by Hermann (1963) as early as the 1960s, and subsequently adopted by a wide range of scholars. First, in a crisis situation, a threat is posed to the essential goals or values of an actor. Second, there is limited time for decision-making, due to an approaching deadline and/or the increasing costs of inactivity. Third, there are numerous unpredictable events or uncertainties connected to the situation, and hence it is difficult to formulate a clear picture of the outcomes of the decisions and actions taken.

This broad definition has several inherent characteristics that frame the discussion in this book, enabling its holistic approach. Above all, the definition is applicable to almost any context in terms of the source of the crisis or the actors involved. Moreover, it has relevance for a range of scenarios, from corporate crises to natural disasters; from the civilian to the military field; from individual decision-makers to the organisational level; and from political leaders to emergency first-responders. While the issue of the typology of crises is important and will be discussed later in this book, the broad definition is applicable to all types of crises, regardless of whether one categorises them according to their causes (e.g. malicious, non-malicious, man-made, technological, natural, social, economic, complex), or their consequences (e.g. social impact, economic impact, human impact, reputation loss) or the degree of the ability to predict and influence them (e.g. Veenema and Woolsey, 2012; Bellavita, 2006; Gundel, 2005; Kruke, 2015).

Additionally, the definition does not presuppose that one defines a crisis in an objective way. It even includes the possibility that a crisis might be a perception rather than an objective reality, or that crises are sometimes, either intentionally or inadvertently, self-constructed by the very actors who are responsible for managing them, and not necessarily created or caused by any objective circumstances. Furthermore, while the broad definition above leads us to focus on crisis decision-making, and emphasises time constraints as one of the main characteristics, it does not limit the timeframe of the crisis itself. Our broad definition covers sudden events such as a plane crash or the assassination of a political leader, but can also be applied to creeping crises, such as climate change or financial bubbles. In other words, it applies to those four types of crises that Boin, McConnell and 't Hart (2008, p. 41) have identified for analytical purpose: (1) fast-burning crises with instant development and abrupt termination; (2) cathartic crises with creeping development and abrupt termination; (3) long-shadow crises with instant development and gradual termination; and (4) slow-burning crises with creeping development and gradual termination.

Added to this, a rather recent discussion (e.g. Schiffino et al., 2017) has revolved around the argument that the nature of crises has changed in tandem with globalisation. Modern crises are, it is claimed, different from crises in the past in that they are non-linear, trans-sectoral and trans-boundary, and associated with unknown origins and remedies. Be that as it may, the broad definition of a crisis outlined above also covers these new crisis dynamics.

The second part of the concept is the management aspect. One can understand management as a set of interrelated or interacting elements to establish policy and objectives and to achieve those objectives (ISO, 2015). This definition comes from the International Organization of Standardization 9000 family of standards that focuses on the quality of management in general. It is particularly apt in relation to crisis management, however, where there is a clear normative purpose by definition, namely better management of crises.

One can also find several definitions of crisis management, typically deriving from international organisations or national government documents, or as a

4 *Introduction: The crisis management cycle*

part of lower-level standards (see CIPedia, n.d.). Basically, one can distinguish between the narrower definitions that put emphasis on the very crisis response, and the broader definitions that emphasise the importance of management before, during and after the crisis. This book relies on the broader understanding. While a crisis is a temporary event, crisis management is an ongoing and indeed essential activity even between crises.

1.2 What is the crisis management cycle?

Following the ISO notion of management touched on above, in a broader sense, crisis management indeed includes many interrelated or interacting elements. However, it is difficult to find a developed theory of crisis management that would draw all these elements together. Somewhat paradoxically, there is a great deal of middle-level theorising about individual elements or themes. One may find several coexisting approaches and sometimes rival models, theories, lively discussion, or even schools of thought or discourses related to risks, crisis decision-making, human factors in crisis management, crisis communication, crisis and media, or learning from crises, and so forth. The reason for this is that these crisis management sub-themes are often rooted in more developed disciplines and theoretical backgrounds. Indeed, most of the material in the increasing number of academic journals dealing with crisis management issues could be classified as belonging to this middle-level theorising approach, often accompanied by single or small-n case studies. While it is impossible to cover all of this literature in one volume, the present book nevertheless makes an effort to at least identify the main trends and debates in the field.

Boin et al. (2005; cf. Stern, 2013, 2014) have argued that crisis management can be usefully broken down into five key challenges, common to any type of crisis and actor. Sensemaking refers to the interpretation of the often-complex crisis situation. Decision-making includes the main decisions or series of decisions that crisis managers have to make during the crisis. Meaning-making basically concerns how leaders communicate the crisis to other stakeholders or society at large. Terminating is the phase when a crisis situation returns to normality. Finally, learning refers the post-crisis process whereby lessons are learned, as well as their institutionalisation. One can also find several other theories or conceptualisations aimed at dissecting crisis management into parts that would enable both better normative guidance and theoretically more analytical treatment (for an overview of some models or theories, see Sellnow, Seeger and Sellnow, 2013, pp. 25–48).

However, one of the most holistic and at the same time more operational understandings of crisis management is the typological-chronological so-called crisis management cycle, also referred to by other names, such as the emergency management cycle or crisis life cycle (e.g. Heath, 1998; Heck, 1991; Rosenthal and Pijnenberg, 1991). If a crisis is understood as a time-limited phenomenon, and a kind of deviation from the normal state of affairs, from the cycle perspective, crisis management is not: it also covers the normal state of affairs before and after the crisis.

The cycle frame provides a very practical way to distinguish between the different phases of this management system. Relying on the cycle frame, one can combine the different levels, dimensions, elements and phases of crisis management literature and practice. In its standard version, the crisis management cycle always includes at least pre-crisis, during-crisis and post-crisis phases. Effectively, this suggests that any successful management of a crisis presupposes a long-term and multidimensional management (or governance) system. In practical terms, the crisis management cycle provides a framework for putting the interrelated or interacting elements into some kind of order and discussing issues such as who should be doing what and when in order to successfully manage a crisis situation. This is why the crisis management cycle has largely been utilised in the context of advising practitioners, in terms of manuals, handbooks and guidelines for action for employees in companies and public administration. However, the cycle approach is also a good starting point for a more theoretical treatment of crisis management, as this book aims to demonstrate.

The pre-crisis, during-crisis and post-crisis phases of the crisis management cycle are usually further divided, with some variations, into more detailed and prescriptive phases. These may include the following phases, which have been used to structure the current book: risk assessment, prevention, preparedness, response, recovery and learning. The cycle based on these phases is illustrated in Figure 1.1.

Effective crisis management presupposes the efficient management of any of its individual phases. A highly convenient starting point is that many of these phases are rather well defined by international standardisation organisations or other international bodies. This is the case particularly in the field of disaster reduction and business continuity, usually for very practical purposes. With certain modifications, these definitions are nevertheless applicable to almost any field of crisis management.

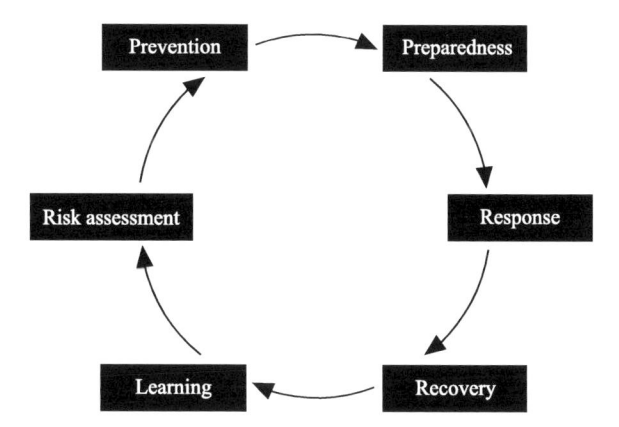

Figure 1.1 The crisis management cycle

1.3 The aim of the book

The current book uses the above-mentioned phases of the crisis management cycle as its building blocks, with each chapter focusing on one phase. It has to be highlighted, however, that even if the crisis management cycle gives the impression of a highly sequential system, it is only a model that is employed to conceptualise the numerous issues discussed within the respective literature and dealt with in practice. While a certain sequence exists, the phases overlap, and in practice they are often carried out simultaneously by organisations or societies, and not wholly demarcated as such.

At first glance, the book provides a rather exhaustive literature review. It has two more far-reaching aims, however. First, almost every phase of the crisis management cycle is characterised by several coexisting approaches, and sometimes internal debates and rival discourses. The current book's task is to identify these different understandings, systematise them, and position them in relation to each other.

Second, by using the crisis management cycle as a framework, the book draws a holistic and generic picture of the field of crisis management. This helps to overcome the conceptual and discursive chaos and lack of theoretical or methodological cohesion that a student of this field ostensibly feels when trying to make sense of the plethora of literature related to crisis management. While the book itself does not develop a holistic theory of crisis management, the concluding chapter discusses the issue of whether a theory of crisis management might be possible and, if so, what such a theory might look like in light of the findings.

This approach provides a rather comprehensive view of what crisis management entails. While each chapter delves into its specific theme in some detail, the main issues dealt with in the book are gathered together and summarised in the concluding chapter. Readers who are keen to gain a quick overview before reading the individual chapters may want to turn to Table 8.1 in the concluding chapter first.

References

Bellavita, C. (2006) Changing Homeland Security: Shape Patterns, Not Programs. *Homeland Security Affairs*, 2(3). Available at: http://www.hsaj.org

Boin, A. et al. (2005) *The Politics of Crisis Management: Public Leadership Under Pressure*. Cambridge: Cambridge University Press.

Boin, A. and Lagadec, P. (2000) Preparing for the Future: Critical Challenges in Crisis Management. *Journal of Contingencies and Crisis Management*, 8(4), pp. 185–191.

Boin, A., McConnell, A. and 't Hart, P. (2008) *Governing After Crisis: The Politics of Investigation, Accountability and Learning*. Cambridge: Cambridge University Press.

CIPedia (n.d.) Available at: https://publicwiki-01.fraunhofer.de/CIPedia/index.php/CIPedia%C2%A9_Main_Page

Crandall, W.R., Parnell, J.A. and Spillan, J.E. (2014) *Crisis Management: Leading in the New Strategy Landscape*. Second edn. Los Angeles: Sage.

Deverell, E., Hansén, D. and Olsson, E-K. (eds) (2015) *Perspektiv på krishantering*. Lund: Studentlitteratur.
Dezenhall, E. and Weber, J. (2011) *Damage Control: The Essential Lessons of Crisis Management*. Westport, CT: Prospecta Press.
Drennan, L., McConnell, A. and Stark, A. (2015) *Risk and Crisis Management in the Public Sector*. Second edn. London: Routledge.
Fagel, M.J. (2013) *Crisis Management and Emergency Planning: Preparing for Today's Challenges*. Boca Raton, FL: CRC Press.
Farazamand, A. (2014) *Crisis and Emergency Management: Theory and Practice*. Second edn. Boca Raton, FL: CRC Press.
Fink, S. (2002) *Crisis Management: Planning for the Inevitable*. Lincoln, NE: iUniverse Inc.
Gundel, S. (2005) Towards a New Typology of Crises. *Journal of Contingencies and Crisis Management*, 13(3), pp. 106–115.
Haddow, G.D., Bullock, J.A. and Coppola, D.P. (2011) *Introduction to Emergency Management*. Fifth edn. Amsterdam: Elsevier.
Harvard Business School (2004) *Crisis Management: Master the Skills to Prevent Disasters*. Boston, MA: Harvard Business School Press.
Heath, R. (1998) Dealing with the Complete Crisis: The Crisis Management Shell Structure. *Safety Science*, 30, pp. 139–150.
Heck, J.P. (1991) Comments on 'The Zeebrugge Ferry Disaster'. In Rosenthal, U. and Pijnenburg, B. (eds) *Crisis Management and Decision Making*. Dordrecht: Kluwer, pp. 75–79.
Hermann, C.F. (1963) Some Consequences of Crises Which Limit the Viability of Organizations. *Administrative Science Quarterly*, 8, pp. 61–82.
ISO (2009a) Risk Management – Principles and Guidelines. ISO 31000:2009. Available at: www.iso.org
ISO (2009b) Risk Management – Vocabulary – Guidelines for Use in Standards. ISO Guide 73:2009. Available at: www.iso.org
ISO (2015) Quality Management Systems – Fundamentals and Vocabulary. ISO 9000:2015. Available at: www.iso.org
ISO/PAS (2007) Societal security – Guideline for Incident Preparedness and Operational Continuity Management. ISO/PAS 22399:2007.
Kruke, B.I. (2015) Planning for Crisis Response: The Case of the Population Contribution. In Podofillini, L. et al. (eds) *Safety and Reliability of Complex Engineered Systems*. London: Routledge.
McEntire, D.A. (2015) *Disaster Response and Recovery: Strategies and Tactics for Resilience*. Hoboken, NJ: John Wiley & Sons.
Quarantelli, E.L. (1996) The Future Is Not the Past Repeated: Projecting Disasters in the 21st Century from Current Trends. *Journal of Contingencies and Crisis Management*, 13(3), pp. 228–240.
Regester, M. and Larkin, J. (2008) *Risk Issues and Crisis Management in Public Relations: A Casebook of Best Practice*. Fourth edn. London: Kogan Page.
Robert, B. and Lajtha, C. (2002) A New Approach to Crisis Management. *Journal of Contingencies and Crisis Management*, 10(4), pp. 181–191.
Rosenthal, U. and Pijnenburg, B. (1991) *Crisis Management and Decision Making: Simulation Oriented Scenarios*. Dordrecht: Kluwer.
Rosenthal, U., Charles, M.T. and 't Hart, P. (1998) *Coping with Crises: The Management of Disasters, Riots, and Terrorism*. Springfield, IL: Charles C. Thomas.

Roux-Dufort, C. and Lalonde, C. (2013) Editorial: Exploring the Theoretical Foundations of Crisis Management. *Journal of Contingencies and Crisis Management*, 21(1), pp. 1–3.

Schiffino, N., Taskin, L., Donis, C. and Raone, J. (2017) Postcrisis Learning in Public Agencies: What Do We Learn from Both Actors and Institutions? *Policy Studies*, 38(1), pp. 59–75.

Sellnow, T.L., Seeger, M.W. and Sellnow, T.L. (2013) *Foundations in Communication Theory: Theorizing Crisis Communication (1)*. Somerset, NJ: Wiley-Blackwell.

Stern, E. (2013) Preparing: The Sixth Task of Crisis Leadership. *Journal of Leadership Studies*, 7(3), pp. 51–56.

Stern, E. (ed.) (2014) *Designing Crisis Management Training and Exercises for Strategic Leaders: A Swedish and United States Collaborative Project*. Stockholm: Swedish National Defense College.

UNDHA (1992) *Internationally Agreed Glossary of Basic Terms Related to Disaster Management*. United Nations Department of Humanitarian Affairs. Available at: http://reliefweb.int/sites/reliefweb.int/files/resources/004DFD3E15 B69A67C1256C4C006225C2-dha-glossary-1992.pdf

UNISDR. (2009) *UNISDR Terminology on Disaster Risk Reduction*. United Nations International Strategy for Disaster Reduction (UNISDR), Geneva, Switzerland, May 2009. Available at: http://www.unisdr.org/we/inform/terminology.

Veenema, T.G. and Woolsey, C. (2012) Essentials of Disaster Planning. In Veenema, T.G. (ed.) *Disaster Nursing and Emergency Preparedness for Chemical, Biological, and Radiological Terrorism and Other Hazards*. Third edn. New York: Springer, pp. 1–20.

Watters, J. (2014) *Disaster Recovery, Crisis Response, and Business Continuity: A Management Desk Reference*. New York: Apress.

Ziek, P. (2015) Crisis vs. Controversy. *Journal of Contingencies and Crisis Management*, 23(1), pp. 36–41.

2 Risk assessment

In mainstream discussions on the crisis management cycle, risk assessment is not usually seen as one of its separate phases, but rather as a part of prevention. Yet risk assessment is important in its own right, as it is indeed the basis for successful crisis management and a precondition for the subsequent phases of the crisis management cycle. Trying to prevent or mitigate a crisis becomes meaningless if one does not know what to prevent or mitigate, which risks should be considered and how they should be prioritised. The same logic is applicable to other segments of the crisis management procedure, such as preparedness or warning systems. We should not prepare for risks randomly, and similarly it should be impossible to build monitoring and warning systems for all possible risks. Even in the response phase, risk assessment provides not only important information about a certain materialised risk but also points to such related risks that – if not adequately understood and reflected in the response strategy – could lead to undesired cascading or side effects. In the recovery and reconstruction phases, risk assessment provides information about the proper allocation of funds. Post-crisis learning, in turn, is the element that feeds into future risk assessment. This last notion reminds us that risk assessment, while looking forwards, also always draws on lessons learned from previous crises, disasters and emergencies. In any event, risk assessment is essential to understanding and dealing with crisis management, and it deserves its own chapter in a book like this.

This chapter will focus overwhelmingly on risk assessment from the perspective of what is associated with the risk management concept. This discourse can be viewed as a rather formalistic one, and it is to a large extent standardised and normative. It is therefore often criticised as representing an overly technological – indeed, engineering – view of risks, and as expecting that most risks can be rationally identified, measured and treated. It is easy to acknowledge that the two partially rival discourses – the traditional risk society discourse and the more recent risk governance discourse – offer a much more holistic picture of risk from the perspective of the respective society at large. The shortcomings and limitations of the normative risk management discourse notwithstanding, it offers good operational tools to manage crises from an organisation's point of view, which is the perspective that is mainly followed, rather than that of the whole society. The chapter also briefly discusses the broader risk society and risk governance discourses.

The chapter is organised into five sections. In the first section, the concepts of risk and risk assessment are elaborated, dividing the latter into its widely acknowledged three phases: risk identification, risk analysis and risk evaluation. The theme of the second section is the first phase of risk assessment – that of risk identification. It explains the importance of data and discusses whether it is possible to identify all risks. The third section presents the two components of risk analysis, namely, likelihood and consequence analysis. In the fourth section, risk evaluation is briefly discussed, elaborating in particular on such issues as risk tolerance and risk perception. Finally, the chapter takes an illustrative look at some of the techniques and methods, both quantitative and qualitative, that are used in mainstream risk assessment. While one cannot draw a comprehensive picture of the multiple techniques in the current book's context – insomuch as there is a great deal of literature specific to that subject – every crisis manager must have a basic understanding of the character of these types of techniques.

2.1 Decision-making under uncertainty

Risk assessment as such has a long history, closely connected to the development of probability calculation. While the precursors of contemporary risk analysis can be traced back as far as early Mesopotamia before the end of the fifteenth century (Covello and Mumpower, 1985), risk events were typically ascribed to the will of God(s). Bernstein (1996, p. 18) has claimed that up to the time of the Renaissance, "people perceived the future as little more than a matter of luck or the result of random variations, and most of their decisions were driven by instinct". Since then, probability estimates have increasingly become a subject for mathematicians, starting from the puzzle of a simple two-coin toss. The methods were gradually developed, providing an impetus to the insurance markets evolving in London in the nineteenth century (Andersen, Garvey and Roggi, 2014, p. 41).

Currently, it is widely understood that risk assessment is needed in any field and organisation, small or large, operating in the public or private sector, and can be applied in the short or long term, in projects, or in entire institutions. In general, risk assessment has many benefits. It provides a risk management instrument for companies, authorities and policy-makers, public interest groups, civil society organisations and other public or private stakeholders. It informs decisions on how to prioritise investments, and it contributes to the raising of organisational or public awareness of risks. Without proper risk assessment, an organisation is doomed not to manage the unavoidable risks in its activities.

Indeed, risk assessment is at the core of industrial facilities as well as business and financial organisations. While some maintain that it is therefore "too important a responsibility for a firm's manager to delegate" (Coleman, 2011,

p. 1), in many cases risk assessment is nevertheless subcontracted. There is a whole service industry of risk assessment, focusing on, to mention but one example, political risks for a company expanding into a foreign market by way of a joint venture or subsidiary. These risks may include confiscation, expropriation, nationalisation, political violence, currency inconvertibility problems and several types of discrimination. When these risks are identified and quantified, one can make correct investment decisions and plan safeguards, such as insurance policies in the event that the risks materialise.

While risk assessment is commonplace in corporations, most governments also carry out risk assessments. For instance, in recent years, in Europe, risk assessment has become more important in such fields as civil protection and emergency planning. Since 2012, most European countries have prepared such a risk assessment at a national level (e.g. MOIF, 2016; DSB, 2014; MSB, 2012b), generally following jointly agreed methodologies and guidelines provided by the European Commission (European Commission, 2010). In order to coordinate related crisis management activities, the European Commission has summarised these national risk assessments into an overall European one, focusing on risks that are shared by many countries (European Commission, 2014).

What is a risk?

The concepts of risk and uncertainty are sometimes used interchangeably, whereas some make a distinction between them. Should one make such a distinction, risk would be based on our knowledge of historical data in the case of recurring events, coupled with a documented loss of history that displays a certain regularity in the way that the risk scenarios play out. Risk is therefore about probabilities, about the cumulative effect of the probability of uncertain positive or negative occurrences. By contrast, we might speak about uncertainties if one is dealing with extreme and unique events that happen with long time intervals, or if outcomes emerge in completely irregular and unpredictable ways. For these types of uncertainties, it is not possible to assign meaningful probabilities. The distinction between risk and uncertainty is therefore that risks can be described as probabilistic outcome distributions, whereas uncertainty cannot (Andersen, Garvey and Roggi, 2014, pp. 37, 38; Pritchard, 2015, p. 7).

While the above-mentioned distinction is perhaps too strict, one could nevertheless conclude from it that one is dealing with risks at various of levels of uncertainty. Another important point is the notion that a risk, like a crisis, is not necessarily negative. As Purdy (2010, p. 882) put it, "It is now widely understood that risk is simply a fact of life and is neither inherently good nor inherently bad." Particularly from the perspective of business and financial risks, it is often noted that "risk is as much about exploiting opportunities for gain as it is about avoiding downside" (Coleman, 2011, p. 12). The International Standardization Organization (ISO) approach towards risk management agrees with this dual understanding of risk as both a threat and an opportunity. In ISO parlance, risk is the "effect of uncertainty on objectives", which can be either negative, positive or both (ISO, 2009b, p. 1). However, in the field of safety and security, a risk is commonly defined as the combination of the probability of a hazard and its negative consequences, where the concept of a hazard is used in the same way as one uses that of a threat in everyday language (UNISDR, 2009).

Let us agree, however, that at the definition level we also keep the potentially positive consequences in mind. The slogan of the founders of Google, namely that a crisis is a terrible thing to waste, is good to keep in mind. The question then largely concerns the probability of these consequences materialising. Instead of the mathematical term 'probability', the term 'likelihood' is also often used synonymously, especially in qualitative discussions on risk. The term 'consequence' is, in turn, often replaced by the quantitative term 'impact'. In any case, this relationship is often expressed in mathematical terms.

$$\text{Risk} = \text{hazard impact} \times \text{probability of occurrence}$$

While the above is valid in nonfunctional relationships, a functional relationship is, instead, a relationship in which the value of one variable varies with changes in the values of a second variable. This means that if the hazard impact is dependent on the probability, or the probability is dependent on the impact, one uses the terms 'vulnerability' (V) and 'exposure' (E). The result is a functional curve, which is expressed as follows.

$$\text{Risk} = f(p \times E \times V)$$

In the European Commission's risk assessment guidelines, for instance, it is explained that using the term 'vulnerability' makes it more explicit that the impacts of a hazard are also a function of the preventive and preparatory measures that are employed to reduce the risk: "For example, for a heat wave hazard it may be the case that behavioural preparedness measures, such as information and advice, can critically reduce the vulnerability of a population to the risk of excess death" (European Commission, 2010, p. 16).

What is risk assessment?

Risk assessment is perhaps the most written-about topic among those related to the subject of this book, including several dedicated journals related to a number of academic disciplines, and several well-written academic and more practical textbooks, manuals and guidelines (e.g. Pritchard, 2015; Jha and Stanton-Geddes, 2013; Yoe, 2012; Ostrom and Wilhelmsen, 2012; MSB, 2012b; Coleman, 2011; Aven, 2008). Even a cursory glance at this literature reveals, however, that the vocabulary of risk assessment is extremely confusing. There seems to be no consensus on what is meant, for instance, by such concepts as risk assessment, risk analysis, risk evaluation and risk management (see Renn, 2006, Annexes A, B, pp. 86–156). Indeed, they are often used with a variety of meanings, and sometimes interchangeably.

While it should not matter at the end of the day whether one uses one approach and the respective vocabulary consistently, it is nevertheless a welcome development that the field has in recent years become an object of authorised standardisation bodies, most notably the ISO. The so-called ISO 31000 family of standards (ISO, 2009a, 2009b; ISO/IEC, 2009) is the core of this standardisation effort. An important feature of the ISO approach, although criticised by some experts and organisations (see Purdy, 2010, p. 885), is that it is aimed at reflecting a multidisciplinary and generic process, unrelated to a specific domain or type of organisation.

According to ISO 31000, the umbrella concept is risk management, which refers to all coordinated activities to direct and control an organisation with regard to risk, as illustrated in Figure 2.1. Indeed, the ISO risk management concept overlaps to some extent with what in the current book is discussed in terms of crisis management, especially with regard to the phases of prevention and preparedness. The ISO standard is the result of the work of a working group representing hundreds of experts from 28 countries, and was adopted almost unanimously with only one country voting against. The standard is also designed to be dynamic, iterative and responsive to change. This is achieved through the ISO practice whereby standards are always subject to regular reviews after a few years. Yet it has to be kept in mind that the ISO 31000

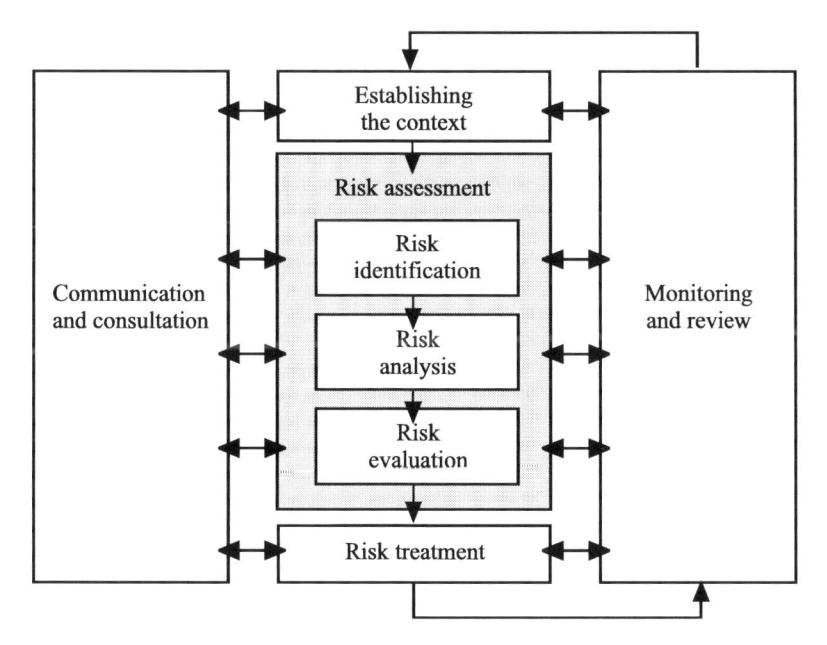

Figure 2.1 ISO 31000 risk management framework

standard family provides only a generic risk assessment (and management) framework, and should be tailored to the specific and diverse needs of the organisation in question.

Risk assessment – the topic of the current chapter – is a part of the broader risk management process. Risk assessment in turn consists of three tasks: risk identification, risk analysis and risk evaluation. Risk identification is the initial process of finding, recognising and recording risks. Risk analysis is about developing an understanding of the risk by determining the consequences and their probabilities for the identified risks. Risk evaluation delineates the significance of the level and type of risk; in this final phase of risk assessment, the risks are prioritised considering organisational goals, regulative requirements, political, ethical, financial and other factors in order to make a balanced decision about future actions.

There has been some criticism of the ISO 31000. It has been argued that the intended meaning of the standard's "rather abstract text is frustratingly hard to pin down", or that the key words and phrases are "either vague, have meanings different from those of ordinary language, or even change their meaning from one place to another" (Leich, 2010, p. 887; cf. Yoe, 2012, p. 230; and Lalonde and Boiral, 2012). Others have been more generous, even stating that the standard represents "a very significant milestone in mankind's journey to understand and harness uncertainty" (Purdy, 2010, p. 886).

This situation is somewhat further obfuscated by the introduction of the rather new concept of risk governance, which is an infiltration of the fashionable governance or multi-level governance concept into the field of risk discourses. Unlike risk management, which is usually understood as focusing on an organisation, risk governance "looks at the complex web of actors, rules, conventions, processes and mechanisms concerned with how relevant risk information is collected, analysed, communicated and how management decisions are taken" (see Renn, 2008, p. 9). Thus, being a much broader concept than management, governance is understood as a complex coordination of different stakeholders. Hierarchically, it is organised from local, regional, national and supranational levels to global levels. Horizontally, it covers actors such as governments, industries, science and academia, and non-governmental organisations. Both axes are enabled and constrained by regulations, markets, norms, culture and different perceptions of risk.

While the governance perspective adds some important aspects to the risk debates, informed in particular by related social science debates, it nevertheless remains rather abstract for our purposes. Moreover, it tends to neglect the fact that the risk management approach also deals with such challenges as uncertainty, ambiguity and complexity. Furthermore, risk management discourse considers stakeholder involvement and communication, and relies heavily on the need to use multidisciplinary approaches and on balancing the pros and cons of precautions in risk-taking. Thus, it would be wrong to introduce these issues as risk governance's own innovations and additions to the risk debates (cf. Van Asselt and Renn, 2011). At the same time, one should be open to alternative understandings that exist in literature and practice, including the risk governance discourse. The standardised ISO approach will be used as the backbone here – not least to avoid further terminological confusion. However, complementary and sometimes even critical dimensions will be purposefully added to it.

The context of risk assessment

Before an organisation embarks on risk identification, several parameters must be in place. The ISO standard speaks about establishing the context (ISO, 2009a), while others may use such expressions as defining the risk environment (Pritchard, 2015, pp. 26–33). In any event, this means that an organisation that is planning to prepare a risk assessment must first articulate its risk management context in terms of its general mission, values and objectives, as well as different external and internal limitations, enablers, perceptions, and so forth.

For instance, if one speaks about companies, for some of them the assumption that "only profits matter is pretty close to the truth because the primary objective of financial firms is to maximize profits", whereas other things, such as jobs, company ranking, environmental values and so on are secondary (Coleman, 2011, p. 12). For public good organisations, such as the government, the police or an environmental agency, the goals and priorities are naturally quite different. The context of risk assessment should be established accordingly.

One can understand this initial process as writing the Terms of Reference for the risk assessment task. This phase should also define the risk criteria, including, for instance, the metrics and time window to be used. When it comes to the latter, it depends on the subject and purpose of the risk assessment. It may cover only a very short time period, such as one project's lifetime. For business planning or national safety and security risk assessment, it may be reasonable to consider mid-term risks of some 5, 10 or 20 years ahead. In long-term strategic planning, where, for instance, governments have to decide upon long-term energy investments, or when they consider the possibility of superhazards for the earth, such as major solar storms or asteroids, one also has to consider at least so-called once in 50-, 100- and 200-year hazards in risk assessment.

Risk criteria should also include classifying risks pertinent to a certain organisation. In companies, one usually differentiates between external risks and internal risks. One could focus, for instance, on strategic risks, operational risks, financial risks, reputational risks, and so forth, also taking into account their combinations and cascading effects. Alternatively, one might choose to focus on external, internal, technical and legal risks, or those originating from people, human error, processes, technology, the external environment, and so on (Andersen, Garvey and Roggi, 2014, p. 47). In the civil protection field, it is customary to differentiate between natural, non-malicious man-made, malicious man-made and multi-hazards, for instance. Multi-hazards can be understood as hazards taking into account simultaneous, cascading, domino and other types of causal and non-causal developments (European Commission, 2014). If the issue at stake is a risk assessment for a limited project, such items as the schedule or customer relationships might constitute important risk areas (cf. Pritchard, 2015, pp. 10–18).

While in the ISO guidelines this context is to be established as the first step before starting the actual risk assessment, it may well happen that in most cases one has to return to the risk criteria in particular after one has formed a better picture of the risks one faces. In any event, the risk criteria become essential when one moves into the last phase of risk assessment – the risk evaluation phase – where one is supposed to choose between those risks that will be treated and those that will be left untreated.

2.2 Risk identification

After establishing the context, risk identification is the first phase of risk assessment. It comprises a screening or mapping exercise, leading to a preliminary long list of risks and their possible combinations. It should include both risks whereby the source(s) can be controlled by an organisation and those that cannot be controlled. While some emphasise that the risks to be identified should be 'realistic' ones (Pritchard, 2015, p. 33), it is argued in the current book that risk identification should be about both data and imagination.

Risk identification is, to start with, the process of pinpointing, recognising and describing risks, based on some kind of data, usually looking backwards at the historical record. While risk identification may appear at first sight to be simple data gathering, it is, however, always a rather complicated exercise. The quantitative data must be combined with qualitative data and methods typically covering expert opinions in a variety of fields, methodologies to recognise weak signals, creative interpretation of the risk factors for identifiable megatrends (e.g. climate change, or some demographic development), or brainstorming about 'imagining the unimaginable'. In essence, risk assessment – and particularly its first phase, risk identification – are usually carried out as teamwork. With the exception of very limited organisations or projects, one person is not usually capable of having all the requisite information and imagination for proper risk identification. Moreover, the cognitive and psychological limitations of an individual necessitate a risk assessment whereby several types of expertise and knowledge are included.

The importance of data

While imagination is important, it is still fair to state that risk identification should be based, as far as possible, on quantitative data. In a sense, the historical, statistical, spatial and similar information about crises in the past, and their consequences, as well as the current conditions and circumstances under which a crisis might take place, can form the baseline to be qualified with qualitative expert opinions and imagination. This data becomes all the more significant in the subsequent – risk analysis – phase, where it is often the basic way to define the likelihood and consequences. The main challenges in data-gathering are posed by the lack of baseline data or its poor quality as well as by missing time series by which change can be recognised. However, when a lack of proper and robust data is identified in the risk assessment process, this can prompt the development of the data-gathering system in order to enhance the quality of future risk assessments.

Let us imagine a municipality that is concerned about the number of injuries among its population, leading to mortality, morbidity, disability and cost. It regards this issue as a creeping crisis, reaching a point where some extraordinary intervention is justified. The basis for effective safety planning is a detailed and reliable injury registration, surveillance and statistical system. Without knowing what causes the injuries, where and why they happen, and who is affected, it does not make much sense to undertake an intervention (Pursiainen, 2007). In many countries, national authorities have injury registration statistics systems, although they are often simply not detailed enough, nor do they take into account the local, very concrete conditions where the injuries occur. Therefore, in addition to following the national requirements, each community should commence its own injury prevention programme by developing its registration and statistics system. Detailed injury registration helps to define the right measures.

In this case, the injury registration and statistics-gathering process should include, at the very least, such indicators as age groups, gender, injury situation (leisure time, organised after-school time, working time, etc.), mechanism of injury (falls, crush injuries, contact with other people, etc.) and place of injury (home, school, traffic, etc.). In order to become a usable preventive instrument, local statistics could in addition include not only the place in general but the exact place, such as a specific traffic junction or children's playground, and so on. Moreover, it is not simply details of the injury in general that should be recorded, but rather the situation as a whole should be documented as concretely as possible, including the specific type of fall, and so on. Furthermore, such statistics should be presented from which one can derive the combinations of the vulnerable groups, environments and situations, and specific needs in each case. Using these statistics in risk identification, one might be able to discern that injuries are especially high, for instance, in the categories of domestic falls, emergency situations involving the elderly and disabled, children cycling to and from school, male youngsters suffering from leisure-time injuries, adult violence against children, male violence against women, and so on. Having tentatively identified these risks as the most frequent ones, one could go on to analyse them further.

Are all risks identifiable?

Yet risk identification does not always succeed. Sometimes it fails because even when the information is available, it is not always recognised as a potential risk. Weak or even clear risk signals may be lost because of the surrounding 'noise'. The obvious problem here is that if a risk is not identified, it will not be included in the next phase of risk assessment and, consequently, in the subsequent phases of crisis management at large. One cannot prevent or mitigate an unidentified risk, nor can one prepare for it or monitor it in order to release timely warnings.

Are all risks identifiable in advance in the first place? Complexity and chaos theories argue that, in many cases, systems have non-linear dynamics where the behaviour of the system is not causally deterministic, but rather random, and where the randomness is caused by the exponential growth of errors and uncertainties. In this spirit, Bellavita (2006; cf. Gundel, 2005; Klinke and Renn, 2002) differentiates between four kinds of spaces, where a crisis or, in his vocabulary, an emergency can take place. 'The known' is a space where cause and effect are understood and predictable. 'The knowable' is a space where cause and effect relationships may be difficult to recognise, but researchers and experts with sufficient time and resources can determine this relationship. 'The complex' is a space where we can understand cause and effect only retrospectively: what appears logical after the fact is but one among many other logical outcomes that could have occurred. Finally, 'the chaotic' is a space so turbulent that cause and effect are (and remain) unknown.

'The complex' in Bellavita's scheme correlates with Perrow's (1984) well-known 'normal accident' theory. Its basic argument is that in a society, based on high-risk technology, it is impossible to avoid accidents. Perrow defines a normal accident as having four characteristics. First, it sends signals, which provide warnings only in retrospect, making prevention difficult. Second, the accident is made possible by multiple design and equipment failures, which are unavoidable since nothing is perfect. Third, there are always some operator errors involved, which also can be understood only retrospectively. Finally, all the three characteristics may have a 'negative synergy', namely, they coincide, which can lead to a major catastrophe. Accidents or even larger catastrophes inevitably will sooner or later take place especially in systems, which are 'tightly coupled'. In today's language, these could be called interdependent systems or system of systems, where several vulnerabilities amplify, producing incomprehensible, unpredictable, unanticipated and unpreventable outcomes.

As mentioned in the introduction to this chapter, the risk governance discourse (Renn, 2008; Van Asselt and Renn, 2011) has broadened the risk debate somewhat. Risk governance can be understood as a multi-level discourse focusing on the interaction between many actors in the event of larger societal risks, whereas the risk management discourse starts from an organisation's perspective. These approaches are not necessarily at odds, as the organisational perspective has to be included in the multi-level governance approach. Perhaps the most useful element of the multi-level governance discourse is the introduction of the notion of systemic crises, reminiscent of the older risk society discourse (Beck, 1992). Systemic risks are, by definition, mostly neglected in perspectives stemming from an organisation's risks. As the followers of Beck have noted, the organisational perspective is too narrow and the issue is about transforming the language of risk from that of calculation to that of mediation (Adam, Beck and van Loon, 2000, p. 2). Thus, risks should always be negotiated between a wider set of stakeholders. According to Van Asselt and Renn (2011, p. 436), the term 'systemic' describes the extent to which a risk is embedded in the larger contexts of societal processes. They therefore argue that systemic risks require "a more holistic approach to hazard identification, risk assessment, and risk management". In particular, the analysis must "focus on interdependencies and ripple and spillover effects that initiate impact cascades between otherwise unrelated risk clusters". While systemic risks cannot be fully understood, in order to understand them as much as possible, Van Asselt and Renn claim one needs to transcend disciplinary boundaries and to involve knowledge and experience beyond the academic.

Yet it is obviously impossible to foresee all potential crises, even if we had all the potential resources and expertise at our disposal. The situation becomes even more complex because stakeholders often disagree about how to perceive a risk. Sometimes there is no unambiguous interpretation about what is known or knowable. In many cases, risk assessment in general is inevitably probabilistic, while the probability characteristics are poorly understood and false warning rates are high (Basher, 2006, p. 2172).

Much, however, can be achieved based on what we already know or could find out. It has been noted that many disaster losses, rather than stemming from unexpected events, are in fact the predictable result of interactions among systems that are known, such as the physical environment, social and demographic conditions, and so on (Mileti, 1999, p. 3). While some crises are impossible or too complex to forecast or prevent, their theoretical possibility can be taken into account in risk assessment in the identification phase.

One can thus conclude that while risk identification is not capable of recognising all possible risks, it can best facilitate the process if it is aware of its own limitations. As Lalonde and Boiral (2012, p. 293) note, one of the key challenges is "to find a balance between the need to take risks into account as thoroughly as possible in order to implement appropriate preventive measures and the awareness that risk management cannot be reduced to planned measures and organisational routines". Hence, a "blend of anticipation and resilience is required".

2.3 Risk analysis

Risk analysis, as the second phase of the risk assessment process, aims at comprehending the nature of risks in more detail, and determining the level of risk in particular. In risk assessment, one usually deals with several risks identified in the risk identification phase. One then assesses individual risks one by one and finally puts them into a comparable perspective in order to gauge the relative risk levels of different hazards. As discussed above, the combination of consequence (impact) and likelihood (probability) of an event forms the basis for the risk assessment. In simple cases, less severe but more frequent events can mathematically create the same risk as a more severe disaster; that is, however, likely to occur only very rarely. The same generic hazard can represent different risk scenarios, but the risk may still receive the same 'value' in the end. Consider the risk of a once-a-year storm causing damage and the risk of a once-in-35-years storm causing much greater damage, and the algorithm defining risk as a function of impact and probability. The risk of a storm causing damage amounting to €10 million and which is likely to occur on average once a year may be regarded as presenting the same risk as a storm causing damage amounting to €350 million, but where we know from past experience that it is likely to occur only once every 35 years (European Commission, 2010, p. 16). Let us qualify this rather interesting function of consequence and likelihood by looking at likelihood analysis and consequence analysis separately.

Likelihood analysis

Likelihood or, in more mathematical terms, probability, can be estimated based on many factors, such as historical data and time series about events of a similar type and scale. The previous phase of risk identification is crucial here as it provides the material basis for the likelihood analysis. Issues such as geographical analysis (location, extent), temporal analysis (frequency, duration) and dimensional analysis (scale, intensity) should be considered when defining the probability of occurrence.

There are several ways of analysing or presenting likelihood. A quantitative scaling is usually preferred, based, for example, on the frequencies of an incident happening, such as daily, monthly, every 6 months, every 10 years, every 50 years, every 100 years, and every 1000 years. Even more developed quantitative approaches then use more complex techniques to express probabilities, utilising historical data, testing, probability techniques, modelling and simulations, and by taking into account several combinations of affecting variables (MSB, 2012a, pp. 44–46). If one does not have this data, then even a simple qualitative description of the probability of an incident is usually not a good method, especially if one considers several risks at the same time, as it provides no real way of comparing or prioritising the identified risks. Therefore, a typical and better qualitative approach is to use scales containing, for instance, five classes, such as 'very low probability', 'low probability', 'medium probability', 'high probability' and 'very high probability'. If one can, for instance, based on expert opinions expressed in surveys, agree on or make an average evaluation of which risks belong to which probability category, one can then in further phases of risk assessment make more informed decisions on which risks should be treated, and in which order. However, this approach is still rather vague and leaves a lot of room for interpretation.

One challenge in likelihood analysis is to take into account the occurrence of cascading effects of events, or events that coincide because they may considerably affect likelihood. One risk may increase because of the occurrence of another risk, or because another kind of event has significantly altered the vulnerability of the system. The dual problem is that as cascading effects, dependencies and interdependencies consist of unlimited combinations, one cannot usually rely on historical data alone, as what will happen will probably follow a very different causal logic from what has happened earlier and, on the other hand, it is difficult to test these dependencies and interdependencies in an operational environment. Consider cyber security and critical infrastructure. While there are some cyber security test facilities in most developed countries, they cannot be used in their current form to fully test security levels because it is not possible to perform actual operational tests with the information and communication technologies connected to actual critical infrastructure. Therefore, in most cases, the only feasible methodology for studying and testing dependencies, interdependencies and cascading effects is to focus on modelling and simulations (cf. Ouyang and Ouyang, 2014).

By definition, the likelihood of any event is also subject to change over time, depending on changes in societal, environmental, demographic, technological or other conditions (e.g. due to globalisation, climate change, increased traffic, erosion of infrastructure and demographics). It duly becomes important to identify and understand the main drivers in order to pick up on accelerating trends in hazard development. Therefore, it is necessary to be clear about the time window for risk assessment, as discussed above. In doing a likelihood analysis, one should always recognise the uncertainty in the underlying evidence. There are methods to this effect, such as sensitivity analysis, the so-called precautionary principle, and also several probabilistic models that could be used. The specific way to do this depends on the type and quality of the data.

Consequence analysis

In the literature, the potential consequences of a risk are often called impacts in order to convey the measurable property of the consequences. These are often divided into impact categories, depending on the organisation, task or angle. In its disaster risk assessment guidelines, the European Commission (2010), for instance, differentiates between three impact categories. First, human impacts refer to the number of affected people, namely, the number of deaths, the number of severely injured or sick people, and the number of permanently displaced people. Second, economic and environmental impacts, commonly defined as the total of costs such as the following: treatment or healthcare; costs of immediate or longer-term emergency measures; costs of restoration of buildings; public transport systems and infrastructure; property and cultural heritage; costs of environmental restoration and other environmental costs (or environmental damage); costs of disruption of economic activity; value of insurance pay-outs; indirect costs on the economy, indirect social costs; and other direct and indirect costs, as relevant. Third, political and social impacts are usually rated on a semi-quantitative scale and may include categories such as public outrage and anxiety, encroachment on territory, infringement of the international position, violation of the democratic system, socio-psychological implications, damage to cultural assets and other factors considered important which cannot be measured in single units, such as certain environmental damage.

Of course, one could always detail the above categories and differentiate between economic and environmental impacts, or political and social impacts. The chosen impact categories depend on the type of hazard and the type of organisation one is talking about. An adapted model of the aforementioned European Commission guidelines is the Norwegian civil protection authority's categories: life and health; nature and culture; economy; societal stability; and democratic values and the capacity to govern (DSB, 2014, pp. 25, 26). In the risk analysis of a project, the critical impact areas may include overall project success, meeting the project schedule, meeting the project budget, meeting the planned objectives and achieving customer satisfaction (e.g. Raz and Michael, 2001). In a technological case concerning static offshore equipment risk analysis, the international certification agency DNV uses three impact or consequence categories. Safety consequence is the same as the above-mentioned human impact, expressed in terms of the potential loss of life of personnel. Economic consequences are expressed in financial terms using appropriate currency units. Environmental consequences are expressed in terms of mass or volume of a pollutant released into the environment, or in financial terms as the cost of cleaning up the spill, including consideration of fines and other compensation (DNV, 2010, pp. 15, 16). In business risk analysis, in turn, one often focuses on financial risks, which can be divided into several categories: market risk, credit risk, liquidity risk, operational risk, other/legal and regulatory risk, business risk, strategic risk, reputational risk, and so forth. Some of these may be difficult to measure with purely quantitative techniques and may need qualitative consideration as well (Coleman, 2011, pp. 124–127). One can also produce analyses for different risk intensities.

In any case, the consequence or impact analysis has to be based on such indicators and methodology that make the comparison between different risks possible. Second, a way must be found to combine the different impact categories into some measurable and comparable total impact. In the above-mentioned DNV example, for instance, all three impact areas with very different metrics are finally transformed into the same ranking scale A–E, which makes it possible to present a holistic multi-impact output of the risk analysis.

In carrying out an impact analysis, it is also important to evaluate the self-protection capabilities that reduce exposure or vulnerability, in order to estimate the real impact correctly. This means that some barriers or controls that mitigate the consequences are usually already in place, and the effect of these controls should be taken into account in the impact analysis. A further issue to be considered is the situational or circumstantial variables affecting the consequences. For instance, in the case of electricity disruption the risk impact may be quite different depending on whether the disruption takes place during the summer or in the midst of harsh winter temperatures.

2.4 Risk evaluation

Following ISO 31000, risk evaluation is the third phase of risk assessment. This is the process of comparing the results of the risk analysis with the risk criteria to determine whether the risk is acceptable or tolerable. The best-known formal technique to establish the level for defining which of the analysed risks should be treated or even terminated is called the As Low As Reasonably Practicable (ALARP) principle. Another well-known principle for setting the tolerance levels is the so-called precautionary principle. While they are highly relevant to risk evaluation, they are also closely related to risk treatment options. We will therefore discuss these principles in some detail in Chapter 3, and confine ourselves here to highlighting the challenges of risk evaluation from a more generic perspective. The rule of thumb is that risk criteria should be based on the organisation's general risk management policy, which in turn should be in harmony with the organisation's generic objectives and respective strategic and operational plans. A company, for instance, may need to balance between economic concerns, reputation, environmental and safety issues, defining whether a certain risk is high, medium or low, and what this would entail. These criteria should also have taken into account the existing mandatory legislation and other mandatory regulations, voluntary standards and benchmarks that often set the limits for what is acceptable and what is desirable. The interests of acknowledged stakeholders should be considered, as well as related socio-economic factors, among others, especially if the risks are such that they concern not only the organisation in question, but also society at large. The problem lies in how to agree on the same risk evaluation, however. Ultimately, the issue becomes that of risk tolerance and risk perception.

Risk tolerance

If one uses the rather common three-point risk classification scale between high, medium and low risks, the evaluation seems a fairly easy task at first glance. A high risk is not acceptable, and further analysis should be performed to provide a better estimate of the risk. If this analysis still shows unacceptable risks, some risk treatment should be introduced to reduce the criticality. A medium risk may be acceptable, but risk controls (barriers, safeguards) should be considered if reasonably practical. Further analysis should also be performed here to provide a better estimate of the risk. When the risk is low, no further risk-reducing measures are usually required (Haugom, Rikheim and Nilsen, 2002).

Some analysts differentiate between acceptance and tolerance. Tolerability does not stand for acceptability, but refers to the willingness to live with a risk due to its possible benefits, but still trusting or demanding that it should be

24 Risk assessment

properly kept under review and reduced if needed. Tolerating a risk is therefore a balancing act. Building a nuclear power plant is a typical example of a tolerated risk. The decision-makers and the respective society have to balance between the safety risk and the benefits, namely, energy and employment, that it entails (Drennan, McConnell and Stark, 2015, p. 82). In practice, it is, however, always a rather difficult task to decide which risks should be tolerated. As Yoe (2012, p. 79) has noted, this task includes several more delicate issues to decide upon. A risk manager can search for the highest possible level of protection against risks. Alternatively, one can be content with a desirable, achievable, practical and implementable, or simply an affordable level. Thus, there are always several risk acceptance or tolerance strategies between which a risk manager has to choose.

Hunter and Fewtrell (2001) have analysed the different approaches towards establishing acceptable risk in the field of *water quality*. Their first approach is that of 'predefined probability', which is just a quantitative estimate of the probability of injury, disease or death in certain circumstances. Another approach is what Hunter and Fewtrell call a 'currently tolerated' approach. This is virtually a de facto approach at any given time, which defines the accepted risk tantamount to acceptable risk. A 'disease burden' approach entails defining the acceptability in terms of falling below an arbitrarily defined level of water quality-related disease in relation to the total disease burden. An 'economic approach' is a simple cost–benefit approach whereby water quality is measured in monetary terms. For instance, weighing the costs of installing a new sewage treatment system against the expected costs of treating illnesses without this new system. Finally, they discuss the 'public acceptance' approach to the risk, which necessitates a political bargaining process. However, they conclude that no definition of acceptability will be acceptable to all stakeholders.

Varying risk perceptions

The ISO 31000 perspective that has formed the backbone of the discussion above has been criticised because of its assertion that risk assessment "belongs to the realm of rational action and scientific certainty, a realm of clear distinctions between safety and danger, truth and falsity, past and future". By contrast, risks should be seen as social constructions, politicised into discursive strategies for change for any group that is willing to use the language of risk (Adam, Beck and van Loon, 2000, pp. 4, 7).

The argument is that there is no simple way to evaluate even quantitatively well-identified and analysed risks, because to some extent this activity is always related to conflicting perceptions, values and interests. Risk assessment is a question of prioritisation based on subjective evaluation. While in some cases risk analysis produces a rather clear-cut risk matrix where one can easily detect those high risks that should be treated, in many practical cases risk evaluation remains difficult. This is naturally so, not only because risk treatment is often very costly, but also because in many cases conflicting interests and risk perceptions do not allow any optimal equilibrium.

In short, "risk means different things to different people" (Regester and Larkin, 2008, p. 22). If the ISO approach examines risk more from an objective point of view (albeit considering stakeholder interests), from what is called a risk governance perspective, risks as such are understood as a social construction more than any objective reality; hence, "risk assessments are 'mental models'" (Renn, 2008, p. 4). Anderson and Felici (2012) have argued that any organisation may comprise many different groups, whose risk perception may differ radically and whose need for and attitude towards system change vary depending on role and environment. This has implications for how risks are perceived, as the dominance of particular groups may emphasise or de-emphasise certain classes of risk.

Even many of those who would not see risk as totally or primarily socially constructed admit that it is often more a matter of perception of reality than some objective reality. Hard scientific facts can easily be ignored, leading either to risk-assertive behaviour against the evidence, or towards risk taking, even if the scientific evidence speaks strongly against it. Such issues as familiarity with the particular risk, or the level of knowledge and understanding of the risk or its consequences play a role (Melchers, 2001, pp. 202, 203). Moreover, the style of information processing plays an important role in how risk is perceived. As reviewed by Ryu and Kim (2015), a number of sociodemographic characteristics (such as individual ability, motivation, age, gender, income) and context factors (such as the source of information, and the quality of the message) influence the choice of a 'heuristic' (intuitive, spontaneous) or a 'systematic' (deliberate, effortful, critical) information processing mode by which these are combined. Accordingly, higher individual ability, information accuracy and income are positively associated with the systematic mode and negatively with the heuristic mode. Further, the quality of the message has a positive impact on the choice of the heuristic mode. Female gender is negatively associated with systematic information processing (ibid., p. 855).

Next, the choice of the information processing mode influences the way risk is perceived. Thus, Ryu and Kim (2015, pp. 852–854) found that an increase in systematic thinking leads to an increase in perceived risk. In addition, they found that three specific factors positively influenced systematic thinking towards increased risk perception: quality of information, motivation and personal ability. By contrast, even though source credibility, the quality of the message and its vividness were positively associated with the heuristic mode of information processing, they did not have a significant impact on risk perception.

It is also evident that there are differences between individuals; some individuals or groups take risks, while others are more risk-averse (e.g. Sheaffer, Bogler and Sarfaty, 2011). Essentially, every person and organisation has a different degree of risk appetite and attitude. "Hence it is difficult to develop universal rules for dealing with risk" (Pritchard, 2015, pp. xxix, 9). Moreover, the so-called prospect theory has shown that it is not only a matter of individual differences, namely that some people are risk-averse, whereas others are risk-tolerant or even risk-seekers. As will be discussed in more detail in Chapter 5 on Response, it is about how the risk is framed (Kahneman and Tversky, 1979). Another influential theory or conceptual framework in the risk perception literature has been the social amplification of risk (Kasperson et al., 1988, p. 988; cf. Renn, 2008, pp. 38–42). The theory enquires why some relatively minor risks, as assessed by technical experts, often elicit strong public concerns and result in substantial impacts upon society and the economy. The developers of the framework go on to argue that hazards interact with psychological, social, institutional and cultural processes in ways that may amplify or attenuate public responses to the risk or risk event. Risk signals are processed by individual and social amplification stations, including the scientists who communicate the risk assessment, the news media, cultural groups, interpersonal networks and others. Key amplification steps can be identified at each stage. These amplified risks lead to behavioural responses, which in turn may result in secondary impacts.

Van Asselt and Renn (2011) propose three principles of risk governance to overcome some of the challenges related to the variety of risk perceptions: communication and inclusion; integration; and reflection. Communication and inclusion refer to exchanges between policy-makers, experts, stakeholders and the general public. The aim should be to instil trust and provide social support for the responsible governance of ambiguous risks, potentially achieving wider involvement and ownership in risk-related decisions. As tools of inclusion, they suggest exercises, roundtables, open forums, negotiated rule-making, mediation or mixed advisory committees, including scientists and stakeholders. Achieving this calls for what they term social learning. Integration, in turn, refers to the need to collect and synthesise all relevant knowledge and experience from various disciplines, including the range of articulations of risk perceptions and values. Reflection means collective (self-)reflection about balancing the pros and cons of the risk, that is, considering the holistic picture and not only a single risk or the immediate results of its treatment, but potentially also identifying the total net value of any risk-related decision.

2.5 Risk assessment techniques and methods

The analysis above has drawn a rather generic picture of the main elements of the concept of risk and the three risk assessment phases. But how should risk assessment be conducted in practice? There are several recognised risk assessment techniques in existence, which are either quantitative, semi-quantitative or qualitative (e.g. Pritchard, 2015; Jha and Stanton-Geddes, 2013; Raspotniga and Opdahl, 2013; Ward et al., 2010; Yoe, 2012; ISO/IEC, 2009; Chapman, 1998). Semi-quantitative techniques can be understood as those where some originally qualitative variables are quantified along a chosen scale. While a risk manager working at a strategic level does not necessarily need to master all of these techniques in detail, especially those quantitative techniques based on mathematical models, one should be able to understand their usability – their possibilities and limitations – in order to consult the experts when needed.

In many cases, the three different phases of risk assessment – risk identification, risk analysis and risk evaluation – may require the use of separate techniques for each phase (ISO/IEC, 2009, Annex A). Others speak more generally about risk management techniques (Pritchard, 2015) or risk assessment methods (Yoe, 2012), which are chosen depending on the task in hand. However, it is usually agreed among the risk management community that no single technique or model can capture all the risks (Potts et al., 2014), and hence several combinations of quantitative, semi-quantitative and qualitative techniques are typically used in different phases of the risk assessment process.

Quantitative techniques: examples

Starting with quantitative techniques, a good illustrative case is the so-called Bow-tie Analysis, which is actually a diagram that combines two separate techniques, namely Fault Tree Analysis (FTA) and Event Tree Analysis (ETA). The former focuses on analysing the causes of a risk factor, and the latter on the consequences (e.g. Ferdous et al., 2013; Mokhtari et al., 2011). A simplified Bow-tie diagram is shown in Figure 2.2.

The FTA, illustrated in Figure 2.3, is a logical and graphical representation that focuses on the causes that may give rise to a critical event, termed the top event in the diagram. If one imagines that this top event is an explosion in a facility, for example, the FTA describes as far as possible the possible cause of this explosion. It breaks the cause down into smaller chains of events, using two types of 'gates', namely an AND gate and an OR gate. The former means that both events under it have to be present in order for the higher-level event to take place, while the latter anticipates that only one of the events will suffice. In other words, the FTA represents the possible pathways towards the top event. This might be discussed qualitatively, but the technique also facilitates

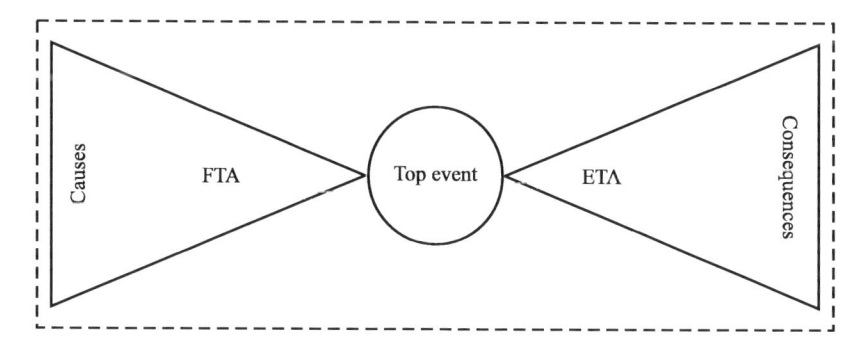

Figure 2.2 Bow-tie diagram

28 *Risk assessment*

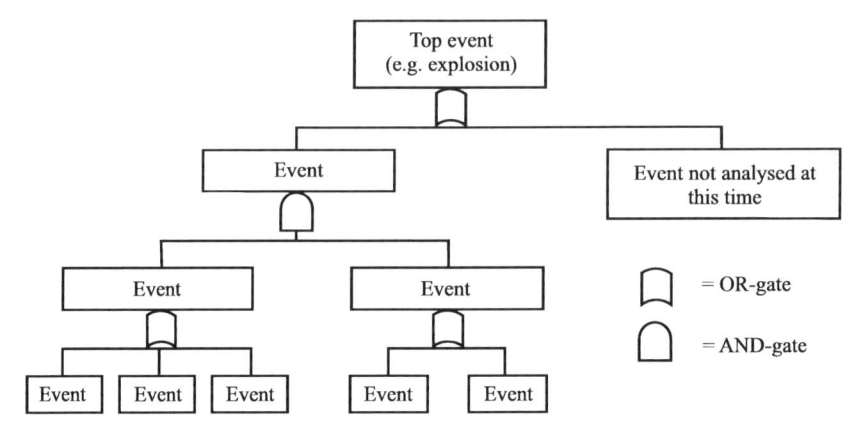

Figure 2.3 Fault Tree Analysis (following ISO 31010, 2009, Annex B)

quantitative analysis. This analysis will then include calculating the probability of each event in the chain. These probabilities can be created, for instance, from historical and statistical data. The final probability of the top event happening can then be aggregated with a normal mathematical algorithm.

Let us now turn to ETA. The technique is presented in Figure 2.4, demonstrating a case where the top event is an explosion once again. Our task is to consider the consequences of the explosion, in this case the risk of fire in the facility.

Figure 2.4 Event Tree Analysis (following ISO 31010, 2009, Annex B)

As Figure 2.4 shows, one starts with the top event. While ETA is best used only qualitatively, the traditional way to use it quantitatively is again to rely on probabilities based on historical data and statistics, performance tests and similar quantitative data. One should, for instance, start by asking how often an explosion takes place per year. Let us imagine that we are speaking about a facility where the probability of an explosion is 10^{-2}, obtained from historical data by calculating the average. The likelihood of the explosion starting a fire is 80% true (yes) and 20% false (no), again based on historical data. The same chain of true-false is then continued with questions as to whether the sprinkler system is functioning, and whether the fire alarm has been activated. Thus, one ends up with five alternative scenarios, the 'uncontrolled fire with no alarm' being the most catastrophic outcome and 'no fire' the least catastrophic outcome, even if the explosion occurred.

While FTA helps us to identify the root causes of the hazard, ETA focuses on consequence analysis. All risk assessment techniques have their pros and cons, however. ETA, for instance, may be appreciated for its ability to describe the sequence of events under different alternative scenarios, but criticised in that it only accounts for binary alternatives, namely, that something happens or not, and there is nothing in between; in our case, one could, for instance, imagine that the sprinklers would only fail partially. This particular technique may also lead to an overly optimistic outlook towards risks, as it does not take into account possible common conditioning simultaneous events that may aggravate the worst-case development of events.

Qualitative techniques: examples

While the first step in risk identification is, in many cases, quantitative, statistical and inductive, that is, expecting something to happen because it has happened before, a deductive and more creative qualitative process is also useful. Quantitative, retrospective data and reliance on formal procedural checking are not enough in risk identification for many reasons. First, as already mentioned, one has to anticipate, if not completely identify, the risks that are yet to come, which may be of a rather different character than past hazards. Usually, however, there are no quantitative, unambiguous metrics or indicators available to quantify such risks. As these risks cannot be overlooked, one has to rely on expert opinions and qualitative estimations. Sometimes one considers so-called emergent risks, that is, those that have not yet occurred but are at an early stage of becoming known or coming into being, and expected to grow considerably in significance (Andersen, Garvey and Roggi, 2014, p. 54). As these risks have never actually materialised, great uncertainty surrounds them. The European Commission, for instance, considers in its risk mapping not only existing but also emerging disaster risks. The latter category includes the impact of climate change and ecosystem degradation, divided into related natural hazards, infrastructure hazards and implications for migration; space environmental hazards, divided into space debris, solar storms and near-Earth objectives; and anti-microbial resistance-related risks (European Commission, 2014). It is also evident that technological innovation will contribute to previously non-existent, emerging risks for any organisation and society at large (Anderson and Felici, 2012).

Second, even if one had identified some of these new risks, one can regard their occurrence as very rare – they may never materialise in our lifetime – but their consequences might be most severe. For instance, the last major solar storm that seriously affected the Earth is the well-known Carrington flare of 1859. The severity follows from the fact that should this extreme Aurora Borealis phenomenon take place today – considering the technology dependency of current societies – many experts estimate that it would, due to its geomagnetic effect, be able to knock out electric power in the (large) affected areas for several months by destroying electrical transmission equipment, especially transformers. In addition, oil and gas pipelines could be affected. Moreover, the space weather risk is also applicable to space-based satellite systems that enable such critical societal tools and services as communication, navigation, meteorology, environment and security monitoring (European Commission, 2014, pp. 57, 58).

Third, an important element even in the risk identification phase is to consider not only single risks, but also complex and multi-risks, where unrelated hazards occur simultaneously, or a cascade of crises and disasters unfolds. Many risks do not follow a sequential linear causality, delineating a certain probability of materialising in such a way as presented above when discussing the Event Tree Analysis, for instance. There are several nonlinear causality chains. Among these, domino causality refers to a situation whereby one risk leads to another, and that risk to another, and so forth. Alternatively, a risk may follow a cyclical or spiralling causality, where the original risk is aggravated by a feedback loop; the second event that occurs due to this feedback may affect the probability of the original risk. Then again, two independent risks may work together to increase each other or cause a third risk. Therefore, in risk identification guidelines one usually advises consideration of both socalled single risks and multi-risks, with expected or unexpected simultaneous, knock-on, domino or cascading effects (European Commission, 2010).

Complex multi-risks are, however, difficult to estimate based on statistics and historical data, because this data does not provide much evidence to anticipate the thousands of potential and unexpected coincidences, dependencies and interdependencies that could take place. Therefore, for these types of new or complicated risks, one may have to add qualitative techniques to identify them. A simple brainstorming session conducted by a small group of key persons, having the necessary knowledge and information based on the shared risk assessment context, may provide a basic understanding of the risks that are imaginable within the limits of the risk criteria. Such brainstorming has to be carefully structured and well led, however, in order to produce a useful risk identification. Expert interviews and self-assessment are also typical qualitative techniques, usually based on semi-structured templates. The former is usually based on external but knowable specialists; the latter is carried out by individuals in the organisation who are asked to identify risks surrounding their area of responsibility. Risk questionnaires and risk surveys can be used by all kinds of target groups, including external and internal stakeholders. A typical SWOT (Strengths, Weaknesses, Opportunities and Threats) analysis is also a good technique for identifying risks, as well as for seeing the positive side of the risks.

Another useful technique is the so-called SWIFT (Structured What If Technique), which can be examined to illustrate a structured brainstorming approach, for example. SWIFT (e.g. Card, Ward and Clarkson, 2012) uses guidewords and prompts to identify, analyse and evaluate risks. It addresses systems and procedures at a rather high level compared to many more quantitative or semi-quantitative approaches. Thus, it simply considers deviations from normal operations with questions beginning with 'What if…?' or 'How could…?'. Checklists to help prevent hazards from being overlooked support this type of brainstorming. While no single standard approach can be associated with this technique, as SWIFT is flexible and modifiable to suit each individual application, it usually starts by defining the systems or processes being analysed, considering each of them sequentially. After that, a group of experts should brainstorm possible hazards. In this phase, one is supposed to merely list the potential hazards, but not discuss them in any detail. The list should then be structured into a logical sequence for discussion, starting with the major ones. Each hazard should be considered one by one, including the possible causes of the event, possible consequences if the event occurs, safeguards (barriers, controls) that are in place to prevent the event from occurring, and its frequency. This discussion should be recorded in the SWIFT log sheets. Finally, one should reconsider whether any hazards have been omitted. When identifying individual hazards, the group should then consider the proper guidewords and prompts. These guidewords are often related to the question of where the hazard originates from. If we were speaking about a certain working community, such as a health clinic or an office, the following guidewords could be considered, for instance: a person, a team, a process, the organisation, the work environment, tasks, technologies and tools, and the external environment.

Let's imagine that we are using SWIFT when pondering the potential risks associated with a nuclear power plant (NPP) located on the coast. Our task is safety and security and our prompt is NATECH, that is, risks related to natural hazard triggered technological disasters (e.g. Krausmann, Gruz and Salzano, 2016). In this case, our structured brainstorming session, utilising SWIFT, might proceed as follows: What if there is a 9.0-magnitude earthquake that cuts off the entire external electricity supply to the NPP? One would conclude that all the NPP's reactors would automatically shut down in the event of the earthquake, as they are designed to do. As a result, however, the reactors would be unable to generate the power to run their own coolant pumps. This would cause the emergency diesel generators to come online, as designed, to run the

32 *Risk assessment*

electronics and cooling systems. But what if the earthquake causes a 15-metre tsunami? One would then notice that the NPP's protective seawall is only 5.7 metres high. In all probability, the tsunami would then destroy the generators for the reactors due to their location in unhardened low-lying areas. One could then go on to ask, what if the problems could not be fixed before the batteries planned as the next redundancy measure run out. This would lead us to the conclusion that the probable consequence would be a reactor meltdown.

While the technique is deceptively simple, and based in part on such maxims as 'ask stupid questions', 'imagine the improbable' and 'think in terms of worst cases with cascading effects', it is nonetheless a rather efficient risk identification technique. The example above shows that only three rather straightforward and nontechnical 'what if' questions would have sufficed in order to identify the risks that basically materialised in the Fukushima Daiichi nuclear disaster in 2011. Had the SWIFT technique been applied in the above fashion, several risk treatment options could have been implemented in advance to limit either the probability of the disaster or to mitigate its consequences.

One should beware of trusting just one technique, however. Focusing on a health clinic, Potts et al. (2014) conducted a study using two risk identification techniques, SWIFT and Healthcare Failure Modes and Effects Analysis (HFMEA). They then compared the risks identified by two separate test groups. The group using SWIFT identified 61 risks, whereas the group using HFMEA identified 72. However, there was only a partial overlap in the results. SWIFT identified 33 risks not identified by HFMEA, and HFMEA identified 42 risks not identified by SWIFT. As this example demonstrates, one should always consider using a mix of qualitative, quantitative and semi-quantitative techniques.

Risk scenarios

The use of scenarios is usually an integral part of any risk assessment because it allows the identified risks to be concretised. Scenarios can be used in all phases of risk assessment. Oftentimes, the result of risk identification is actually a set of scenarios, which are then analysed in the next phase in terms of likelihood and consequences. Scenarios are qualitative and descriptive models of how an identified risk might materialise. While we may know the number of forest fires or industrial accidents in a certain area, or the bottlenecks encountered by a business, only more detailed scenarios will make this information useful in practical terms for comprehensive risk assessment purposes.

While being a qualitative risk analysis method, scenario-building can also be seen as a strategic planning method that combines known facts about the present and the future – such as time, location, socio-economic and other characteristics, residual but plausible factors – with key risk factors. Scenario-building is particularly useful for examining complex developments. It is usually only through scenarios than one can combine many risk factors in different ways to create some surprising events – one of the main characteristics of a crisis – that are difficult to formalise, but which nonetheless better simulate the nature of a real-life crisis.

In practice, risk scenarios are often constructed with certain levels of impact in mind, which means that the consequences are not actually revealed, making scenarios a tool for circular argumentation to a certain extent. However, scenarios are also very useful for concretising and analysing consequences. As for establishing probabilities, scenario analysis alone is rather poor, and often has to rely on other, quantitative techniques. The probabilities of scenarios are then also often selected *a priori* to include hazards exceeding a certain threshold level in order to make the scenario interesting for practical crisis and risk management purposes. However, scenario analysis may help to reveal and analyse the dependencies and interdependencies related to the situation that would otherwise remain unnoticed if one were to rely solely on quantitative techniques.

There are several ways to set certain criteria for scenarios. Scenarios can be used as benchmarking situations that reflect the 'best case', 'worst case' and 'expected case' with regard to a hazard. Alternatively, one can try to vary the conditions of a basic scenario in order to cover a greater number of possible developments. For example, if one ponders civil emergencies, one might consider fluctuating weather conditions, geographical locations, seasons, weekdays and holidays, warning time, technical infrastructure functions, the contributions from other actors, implemented preventive and preparatory measures, consequential events, accessibility, population density in the affected area, expected duration of the event, and so on, depending on the risk and the chosen perspective. In this way, one may examine whether the crisis management capacity applies to a large number of scenarios (MSB, 2012a, pp. 50, 56). This basic idea is illustrated in Table 2.1.

Usually, the basic alternatives are to either consider typical risk scenarios, based on historical experience and statistics or, alternatively, to consider worst-case scenarios that have never materialised. A combination, selectively utilising elements from historical cases and adding some new elements and developments to the scenario, is perhaps the best option from the perspective of prevention and preparedness planning. Yet the scenario-building has to take into account the risk criteria set for risk assessment in advance, such as the time window. In its national risk analysis, the Norwegian civil protection

Table 2.1 Risk scenario with varying conditions

Same risk with three scenarios	Probability per year	Impact category 1	Impact category 2	Impact category 3
Scenario 1	0.1	Low	Low	Medium
Scenario 2	0.01	Medium	Medium	High
Scenario 3	0.001	Medium	High	High

(MSB, 2012b, pp. 50, 56).

authority, for instance, relies on conceivable worst-case scenarios, which could ostensibly occur in the course of a year (DSB, 2014).

The use of short-term scenarios (sometimes called event scenarios) is limited to a situational analysis whereby the context in which a hazard could materialise is constructed imagining the conditions and variables, chain of developments and the probable direct and indirect consequences. In scenario parlance, one prepares a narrative description or a storyline for the scenario. Scenarios should be based on a coherent and internally consistent set of assumptions about key relationships and driving forces. In this sense, a scenario is a case study about a case that has not happened. Instead of inserting an abstract risk into the risk matrix (to be discussed below), one can duly insert the risk scenario instead (cf. DSB, 2014). This type of analysis can be of considerable help in understanding the severity of the hazard as one has to think concretely about the possibilities for dealing with the situation with existing resources.

Long-term scenarios, sometimes called strategic scenarios, are, by nature, rather abstract. In these scenarios, the aspects considered are not situational variables, but rather such factors as external changes, namely, technological changes, decisions that need to be made in the near future but which may have a variety of outcomes, or stakeholder needs and how they might change. Moreover, changes in the macro-environment, including such factors as, for instance, regulatory changes, demographics, climate change, way of life, and so forth, are taken into account and considered from the point of view of the organisation preparing the scenarios. Such macro-environment developments are often understood as macro-drivers or megatrends, some of which are virtually inevitable and some uncertain (ISO/IEC, 2009, p. 42).

Just like any technique, scenario analysis has its benefits and limitations. The benefits include many of the issues already discussed above, that is, concretising abstract risks into something more understandable or even measurable in terms of consequences. The limitations of this technique, especially if it is used as a decision-making tool in too strict a sense, are that the data used in scenario-building may be very speculative, and hence unrealistic results may not be recognised as such (ISO/IEC, 2009, p. 42).

Risk matrix

As defined above, risk is a combination of consequence (impact) and likelihood (probability). This function is often expressed and illustrated in terms of a risk matrix using these two input variables. In ISO language, this is called a consequence/probability matrix, and is a means of combining semi-quantitative ratings of the impact and probability of several risks into the same figure in a comparative perspective. While the risk matrix is a risk analysis technique in its own right, it is also a powerful summary screening and communication tool for the results from the performed risk analysis. A simplified risk matrix, combining three categories of risk impacts (human, economic and environmental, and political-social) is illustrated in Table 2.2.

Table 2.2 Risk matrix

Likelihood	5	Medium	Medium	High	Very high	Very high
	4	Medium	Medium	High	High	Very high
	3	Medium	Medium	High	High	High
	2	Low	Low	Medium	Medium	Medium
	1	Low	Low	Low	Medium	Medium
	Human impacts	#	#	#	#	#
	Economic and environmental impacts	€	€	€	€	€
	Political-social impacts	1	2	3	4	5
		Consequences				

Setting the metrics and designing the visualisation and verbal presentation – such as highlighting low vs. high risks – have to be carefully tailored according to the available data and presentation purpose. One disadvantage of the risk matrix method is that, without any aggregation system that would convert the metrics of different types of impact or consequence factors – say, human impacts (e.g. casualties) and political-social impacts (e.g. political turmoil, public order) – into the same metrics, it is difficult to use only one generic risk matrix for the set of analysed risks. Therefore, it might be necessary to create a separate risk matrix for each impact or consequence category, which in turn will create problems when evaluating the risks that should be treated and those that could be left untreated. On the other hand, this may provide an impetus to develop more tailored risk treatment strategies focusing on particular impact categories, especially if the risk level seems to vary according to the impact category.

Other, more fundamental challenges are associated with the use of a risk matrix. Cox (2008), for one, has correctly noted that the risk matrix has several limitations. Categorisations of severity for uncertain consequences cannot be made objectively, and the resulting outputs of risk analysis require subjective interpretation. Risk matrices should therefore be used with caution. Similarly, Huihui, An and Ning (2010) have argued that the traditional risk matrix, such as the one presented in Table 2.2, suffers from non-meticulous classification of the risk index and the subjective calculation of logic implications. This lack of accuracy leads to a situation whereby different inputs cause the same risk to fall into the same category, so-called risk ties, which makes decision-making difficult.

While this can be avoided to some extent by adding risk matrix cells to emphasise these differences, the sharp boundaries of cells may, however, lead to situations where a very small difference in probability, for instance, will place two risks in completely different risk categories, which cannot be justifiable from a common-sense point of view. This can be partially overcome by introducing a fuzzy risk matrix, that is, including linguistic terms of variation, description range, and so forth. Huihui, An and Ning (2010) present some arithmetic extensions of the risk matrix, which, they claim, solve the problems. These extensions, however, become rather complicated risk assessment techniques in their own right and defeat the illustrative and communicative purpose of the risk matrix in popular use, namely, to illustrate complex issues with one figure. The problems of the risk matrix notwithstanding, it remains a very powerful visualisation or sense-making tool, if accompanied by careful explanations of embedded judgments.

References

Adam, B., Beck, U. and van Loon, J. (eds) (2000) *Risk Society and Beyond: Critical Issues for Social Theory*. London: SAGE Publications Inc.

Andersen, T.J., Garvey, M. and Roggi, O. (2014) *Managing Risk and Opportunity: The Governance of Strategic Risk-Taking*. Oxford: Oxford University Press.

Anderson, S. and Felici, M. (2012) *Emerging Technological Risk: Underpinning the Risk of Technology Innovation*. London: Springer.

Aven, T. (2008) *Risk Analysis: Assessing Uncertainties beyond Expected Values and Probabilities*. Chichester: John Wiley & Sons, Ltd.

Basher, R. (2006) Global Early Warning Systems for Natural Hazards: Systematic and People-Centred. *Philosophical Transactions of the Royal Society*, 364, pp. 2167–2182.

Beck, U. (1992) *Risk Society: Towards a New Modernity*. London: Sage Publications.

Bellavita, C. (2006) Changing Homeland Security: Shape Patterns, Not Programs. *Homeland Security Affairs*, 2(3). Available at: http://www.hsaj.org

Bernstein, P. (1996) *Against the Gods: The Remarkable Story of Risk*. New York: John Wiley & Sons.

Card, A.J., Ward, J. and Clarkson, P.J. (2012) Beyond FMEA: The Structured What-If Technique (SWIFT). *Journal of Healthcare Risk Management*, 31(4), pp. 23–29.

Chapman, R.J. (1998) The Effectiveness of Working Group Risk Identification and Assessment Techniques. *International Journal of Project Management*, 16(6), pp. 333–343.

Coleman, T.S. (2011) *A Practical Guide to Risk Management*. Middletown, DE: CFA Institute.

Covello, V. and Mumpower, J. (1985) Risk Analysis and Risk Management: An Historical Perspective. *Risk Analysis*, 5(2), pp. 103–120.

Cox, A.L (2008) What's Wrong with Risk Matrices? *Risk Analysis*, 28, pp. 497–512.

Cruz, A.M., Steinberg, L.J., Vetere Arallano, A.L., Nordvik, J.-P. and Pisano, F. (2004) *State of the Art in NATECH Risk Management*. European Commission, Joint Research Centre & United Nations International Strategy for Disaster Reduction.

DNV. (2010) *Recommended Practice DNV-RP-G101: Risk Based Inspection of Offshore Topsides Static Mechanical Equipment*. Baerum, Norway: Det Norske Veritas.

Drennan, L., McConnell, A. and Stark, A. (2015) *Risk and Crisis Management in the Public Sector*. Second edn. New York: Routledge.

DSB. (2014). Disasters that May Affect the Norwegian Society. *National Risk Analysis*. Oslo: Norwegian Directorate for Civil Protection.

European Commission. (2010) Commission Staff Working Paper. Risk Assessment and Mapping Guidelines for Disaster Management. Brussels, SEC 1626 final.

European Commission. (2014) Overview of Natural and Man-Made Disaster Risks in the EU. Commission Staff Working Document. Accompanying the document, *Communication from the Commission to the European Parliament, the Council, the European Economic and Social Committee and the Committee of the Regions. The Post-2015 Hyogo Framework for Action: Managing Risks to Achieve Resilience.* Brussels, SWD 134 final.

Ferdous, R., Khan, F., Sadiq, R., Amyotte, P. and Veitch, B. (2013) Analyzing System Safety and Risks Under Uncertainty Using a Bow-Tie Diagram: An Innovative Approach. *Process Safety and Environmental Protection*, 91(1–2), pp. 1–18.

Gundel, S. (2005) Towards a New Typology of Crises. *Journal of Contingencies and Crisis Management*, 13(3), pp. 106–115.

Haugom, G.P., Rikheim, H. and Nilsen, S. (2002) Hydrogen Applications. Risk Acceptance Criteria and Risk Assessment Methodology. Available from: http://www.eihp.org/public/Reports/Final_Report/Sub-Task_Reports/ST5.2/EHEC%20paper_final.pdf.

Huihui, N., An, C. and Ning, C. (2010) Some Extensions on Risk Matrix Approach. *Safety Science*, 48, pp. 1269–1278.

Hunter, P.R. and Fewtrell, L. (2001) Acceptable Risk. In Fewtrell, L. and Bartram, J. (eds) *Water Quality: Guidelines, Standards and Health*. World Health Organization. London: IWA Publishing, pp. 207–227.

ISO. (2009a) Risk Management – Principles and Guidelines. ISO 31000:2009.

ISO. (2009b) Risk Management – Vocabulary – Guidelines for Use in Standards. ISO Guide 73:2009.

ISO/IEC. (2009) Risk Management – Risk Assessment Techniques. IEC/FDIS 31010.

Jha, A.K. and Stanton-Geddes, Z. (2013) *Strong, Safe, and Resilient: A Strategic Policy Guide for Disaster Management in East Asia and the Pacific*. Washington, DC: The World Bank.

Kahneman, D. and Tversky, A. (1979) Prospect Theory: An Analysis of Decision under Risk. *Econometrica*, 47, pp. 263–291.

Kasperson, R.E., Renn, O., Slovic, P., Brown, H.S., Emel, J., Goble, R., Kasperson, J.X. and Ratick, S. (1988) The Social Amplification of Risk: A Conceptual Framework. *Risk Analysis*, 8(2), pp. 177–187.

Klinke, A. and Renn, O. (2002) A New Approach to Risk Evaluation and Management: Risk-Based, Precaution-Based, and Discourse-Based Strategies. *Risk Analysis*, 22(6). pp. 1071–1094.

Krausmann, E., Cruz, A. and Salzano, E. (2016) *NATECH Risk Assessment and Management: Reducing the Risk of Natural-Hazard Impact on Hazardous Installations.* Amsterdam: Elsevier.

Lalonde, C. and Boiral, O. (2012) Managing Risks Through ISO 31000: A Critical Analysis. *Risk Management*, 14, pp. 272–300.

Leich, M. (2010) ISO 31000:2009 – The New International Standard on Risk Management. *Risk Analysis*, 30(6), pp. 887–892.

Melchers, R.E. (2001) On the ALARP Approach to Risk Management. *Reliability Engineering and System Safety*, 71, pp. 201–208.

Mileti, D. (1999) *Disasters by Design: A Reassessment of Natural Hazards in the United States.* Washington, DC: Joseph Henry Press.

MOIF. (2016) *National Risk Assessment 2015. [Finland] Internal Security.* Ministry of the Interior Publication. Helsinki: Ministry of the Interior.

Mokhtari, K., Ren, J., Roberts, C. and Wang, J. (2011) Application of a Generic Bow-Tie Based Risk Analysis Framework on Risk Management of Sea Ports and Offshore Terminals. *Journal of Hazardous Materials*, 192(2), pp. 465–475.

MSB. (2012a) *Guide to Risk and Vulnerability Analyses*. Swedish Civil Contingencies Agency (MSB). Editors Jonas Eriksson and Anna-Karin Juhl. Publ.nr MSB366. DanagårdLiTHO.

MSB. (2012b) *Swedish National Risk Assessment 2012*. Swedish Civil Contingencies Agency (MSB).

Ostrom, L. and Wilhelmsen, C.A. (2012) *Risk Assessment: Tools, Techniques, and Their Applications*, First edn. Hoboken, NJ: John Wiley & Sons, Inc.

Ouyang, M. and Ouyang, M. (2014) Review on Modeling and Simulation of Interdependent Critical Infrastructure Systems. *Reliability Engineering & System Safety*, 121, pp. 43–60.

Perrow, C. (1984) *Normal Accidents: Living with High-Risk Technologies*. New York: Basic Books.

Potts, H.W.W., Anderson, J.E., Colligan, L., Leach, P., Davis, S. and Berman, J. (2014) Assessing the Validity of Prospective Hazard Analysis Methods: A Comparison of Two Techniques. *BMC Health Services Research*, 14(41), pp. 1–10.

Pritchard, C.L. (2015) *Risk Management: Concepts and Guidance*. Fifth edn. Boca Raton, FL: CRC Press.

Purdy, G. (2010) ISO 31000:2009 – Setting a New Standard for Risk Management. *Risk Analysis*, 30(6), pp. 881–886.

Pursiainen, C. (2007) *Does Your Community Have an Injury Prevention Programme? Eurobaltic Guidelines for the Baltic Sea Region*. Stockholm: Nordregio and Swedish Rescue Services Agency.

Raspotniga, C. and Opdahl, A. (2013) Comparing Risk Identification Techniques for Safety and Security Requirements. *The Journal of Systems and Software*, 86, pp. 1124–1151.

Raz, T. and Michael, E. (2001) Use and Benefits of Tools for Project Risk Management. *International Journal of Project Management*, 19(1), pp. 9–17.

Regester, M. and Larkin, J. (2008) *Risk Issues and Crisis Management in Public Relations: A Casebook of Best Practice*. Fourth edn. London: Kogan Page.

Renn, O. (with Annexes by Graham, P.) (2006) *Risk Governance: Towards an Integrative Approach*. Geneva: IRGC.

Renn, O. (2008) *Risk Governance: Coping with Uncertainty in a Complex World*. London: Earthscan.

Ryu, Y. and Kim, S. (2015) Testing the heuristic/systematic information-processing model (HSM) on the perception of risk after the Fukushima nuclear accidents. *Journal of Risk Research*, 18(7), 840–859.

Sheaffer, Z., Bogler, B. and Sarfaty, S. (2011) Leadership Attributes, Masculinity and Risk Taking as Predictors of Crisis Proneness. *Gender in Management: An International Journal*, 26(2), pp. 163–187.

UNISDR. (2009) UNISDR Terminology on Disaster Risk Reduction. United Nations International Strategy for Disaster Reduction (UNISDR). Geneva, Switzerland, May 2009. Available from: http://www.unisdr.org/we/inform/terminology.

Van Asselt, M.B.A. and Renn, O. (2011) Risk Governance. *Journal of Risk Research*, 14(4), pp. 431–449.

Ward, J., Clarkson, P.J., Buckle, P.W. and Jun, G.T. (2010) Prospective Hazard Analysis: Tailoring Prospective Methods to A Healthcare Context. Technical report. Cambridge: Engineering Design Centre, Department of Engineering, University of Cambridge.

Yoe, C. (2012) *Primer on Risk Analysis: Decision Making under Uncertainty*. Boca Raton, FL: CRC Press.

3 Prevention

A rather generic International Organization for Standardization (ISO) definition states that the term prevention consists of measures that enable an organisation to avoid, preclude or limit the impact of a disruption (ISO/PAS, 2007). In relation to natural, man-made and technological hazards, the United Nations International Strategy for Disaster Reduction (UNISDR, 2009) in turn sees prevention as the outright avoidance of the adverse impacts of hazards and related disasters, but because the complete avoidance of losses is not always feasible, the task very often reverts to that of mitigation. That is why, in general usage, the terms prevention and mitigation are usually grouped together, or sometimes even used interchangeably. In the current chapter, we will include in our discussion those mitigation efforts that are made before the crisis event or outbreak.

There are, however, many other related terms and concepts that can be understood as rival discourses of sorts when compared to prevention and mitigation. A term very similar to those of crisis prevention and mitigation is that of risk reduction. Indeed, in disaster or emergency management in particular, disaster (risk) reduction has become such a familiar term that only the abbreviation DRR is used. According to UNISDR (2009), which embraces the term in its name, reduction refers to the concept and practice of reducing disaster risks through systematic efforts to analyse and manage the causal factors of disasters, including through reduced exposure to hazards, lessened vulnerability of people and property, wise management of land and the environment, and improved preparedness for adverse events.

Risk reduction, in turn, comes close to what in ISO 31000 risk management parlance is referred to as risk treatment (ISO, 2009a, 2009b; ISO/IEC, 2009). In the risk management context, risk treatment is the phase that follows immediately after the three-phase risk assessment (risk identification, risk analysis, risk evaluation). Thus, after we have identified and analysed the risks, we evaluate them in terms of what will be treated or left untreated, and then select the proper risk treatment methodology or technique. We consider that the risk treatment discourse again presents the most systematic and generic structure for a discussion of prevention and mitigation issues, at least from the perspective of an organisation that is interested in crisis or risk management. It namely offers a kind of taxonomy that divides the theme(s) of prevention and mitigation into subthemes, or risk treatment options, which help to operationalise the respective strategies and actions. Hence, we will use the risk treatment options as a framework for our discussion here.

Divided into six sections, we will, however, begin the discussion with a slightly different discourse, namely, one focusing on an organisational or societal culture aimed at preventing and mitigating crises, often discussed in terms of a safety or security culture. The essence of this discourse is that prevention and pre-crisis mitigation are not only one-time actions, but have to be embedded in the organisational culture and practices. Thus, this discourse on organisational culture is by no means antagonistic to risk treatment discourse or other prevention and mitigation discourses, but rather constitutes a foundation of sorts. The following five sections discuss the different ISO 31000 risk treatment options, combining some of them under same section: risk avoidance and risk removal; reducing the likelihood or consequences of a risk; risk sharing; risk retaining; and, finally, risk taking as a prevention strategy.

3.1 Prevention as an organisational culture

If an organisation is aiming at establishing a successful crisis prevention and mitigation policy, the starting point, and at the same time the objective, are to create a culture that underlies this policy. In the field of emergency management, this is termed a safety culture. In other fields and contexts, it can be labelled differently depending on the organisation's idiosyncrasies and respective risk perceptions and hazard sources.

The origins of safety culture

While the term safety climate had been in existence for some time (Zohar, 1980), the term safety culture was not coined until the aftermath of the 1986 Chernobyl nuclear power plant accident in the USSR, and first appeared in print in an INSAG (1988) report. INSAG stands for the International Nuclear Safety Group, and was established under the International Atomic Energy Agency (IAEA), specifically created to investigate the causes of the Chernobyl disaster. To put it briefly, the group's first assessments and respective conclusions mirrored those officially issued by the Soviet Union. According to the latter, the disaster took place when the personnel were testing one of the reactors, and the fault was therefore clearly that of the operators, who disconnected a series of technical protection systems. In so doing, they breached the most important operational safety provisions in order to conduct the test. After a while, however, INSAG added another main contributing cause, namely, the design of the reactor. More issues, such as deficiencies in the regulatory regime and cross-sectoral communication, were soon highlighted.

Finally, the group came to the conclusion that it was the overall inadequacy of the safety culture that gave rise to all the errors and caused the disaster. In 1992, it was argued that "safety culture was lacking in the operating regime at Chernobyl" and "safety culture had not been properly instilled in nuclear power plants in the USSR prior to the Chernobyl accident". While many of the prerequisites for a safety culture existed in the regulations, "these were not enforced" and many "other necessary features did not exist at all" (INSAG, 1992, pp. 21, 22).

By 1991, INSAG had already published a whole report on the concept that went beyond this particular disaster. The report was designed to provide guidance on the enhancement of safety culture in the nuclear power production field. It offers the following generic definition of safety culture within the nuclear power plant context: "Safety Culture is that assembly of characteristics and attitudes in organizations and individuals which establishes that, as an overriding priority, nuclear plant safety issues receive the attention warranted by their significance." It further states that the concept involves elements such as individual awareness, knowledge and competence of personnel, commitment at senior management level, both rewards and sanctions, audit and review practices, with readiness to respond to individuals' questioning attitudes, as well as responsibility through the formal assignment and description of duties (INSAG, 1991, pp. 4, 5).

Since then, the concept has become widely popularised outside the nuclear field, and has indeed become commonplace, especially in industrial safety practices and safety literature. While the operationalisations of the concept of safety culture differ (Guldenmund, 2000, 2010; Choudhry, Fang and Mohamed, 2007), the one point on which the literature agrees is that awareness and commitment are the key to a positive safety culture. Organisations with a positive safety culture are characterised by communications founded on mutual trust, by a shared perception of the importance of safety and by confidence in the efficacy of preventive measures (Lee, 1996). Safety culture is something that is shared by groups of people; it reflects the attitudes, beliefs, perceptions and values that employees share in relation to safety (Cox and Cox, 1991). It is often considered that safety culture consists of demands and implementation at several levels of an organisation, most notably at the policy level, managerial level and individual level. The requirements are level-specific, but include respective knowledge and competence, commitment, motivation, supervision, individual awareness and responsibility (INSAG, 1991).

It is often highlighted that a positive safety culture demands top management commitment in particular (Zhu et al., 2016; Wu, Lin and Shiau, 2010). This would then lead to mutual trust and credibility between management and employees, continuous monitoring, system review, and continual improvements in corrective and preventive actions and arrangements (Choudhry, Fang and Mohamed, 2007). However, some researchers question the top-down-led collective pathos in the safety culture literature and argue that "the motor that drives the system to its desirable end will always be particular idealistic individuals, not the system alone or the convictions it promulgates"

(Guldenmund, 2010, p. 1478). It has also been noted that within an organisation, even if committed to a safety or security culture at a general level, there might exist several subcultures with rival risk perceptions and considerable variation in their degrees of commitment (Blazsin and Guldenmund, 2015).

Organisational types

How can one duly ascertain whether an organisation has a positive safety or security culture, aimed at preventing unwanted events and bigger crises? Safety culture is traditionally measured by qualitative observation and questionnaires, sometimes adding quantitative elements to this observation (see e.g. Guldenmund, 2010; Parker, Lawrie and Hudson, 2006; Zohar, 2000). The idea of measurement is normative, and is presumed to enhance a positive safety culture. Management commitment can be promoted by allocating resources and time to inspections, by participating in risk assessments and consultative committee meetings, and by completing actions. Employee involvement can be promoted by aiming at changing individual attitudes and behaviours through verbal instructions, training and warning signage, but even more so by enhancing employees' involvement, ownership and commitment. Promotional strategies may involve mission statements, slogans and logos, published materials such as statistics newsletters or the use of media (Vecchio-Sudus and Griffiths, 2004). If we accept that such actions exert a positive effect, it is possible to measure the extent to which they have been applied.

Yet safety culture ultimately boils down to the different responses that organisations demonstrate towards safety or security concerns. One of the first studies to categorise organisations from this perspective was conducted by Westrum (1996; cf. 2004), focusing on aviation and identifying the different cultures according to the information flow within the organisations. He duly distinguished between three cultures: (a) What he termed the pathological organisation, where information is hidden, messengers who express safety or security concerns are punished, responsibilities are shirked, failure is covered up and new ideas are actively crushed. (b) The bureaucratic organisation, which is somewhat more open. Information is not actively concealed, but may be ignored, messengers are tolerated at best, and responsibility is compartmentalised. The organisation is just and merciful when it comes to failures, but new ideas are seen as creating problems. (c) The generative organisation, in turn, is one where information is actively sought, messengers are trained, responsibilities are shared, failure prompts inquiry and new ideas are welcomed. This original typology has since been accentuated and deepened with the addition of more organisational types (Reason, 1997; Parker, Lawrie and Hudson, 2006).

It has been noted (Parker, Lawrie and Hudson, 2006, p. 554) that Westrum's generative organisation type is reminiscent of the so-called high reliability organisation (HRO), a term coined by Karl Weick (1987; cf. Weick, Sutcliffe and Obstfeld, 1999). An HRO is often exemplified by such branches as air traffic control or the nuclear power industry. These types of organisations with high perceived risks are usually characterised by an ideal combination of centralisation and decentralisation. Compliance with safety and security rules is ensured without surveillance because members of the organisation are intrinsically motivated through a highly developed and mature safety culture. The implicit idea is that HRO practices should be applied to any organisation that takes crisis management, and especially prevention, seriously.

Lagadec (1997), in turn, has identified the fundamental debate between the HRO theorists and the so-called "normal accidents theorists" (see Perrow, 1984, 1994; cf. La Porte, 1994). The former, unlike HRO theorists, resemble the risk society theorists discussed in the previous chapter to some extent, and question whether a safety culture is possible in complex organisations. Rather, they argue, it is illusory to think that they can avoid accidents because tightly-coupled technological systems and complexity make unexpected interactions and failures inevitable. Indeed, while the normal accident school of thought is not new, similar arguments have recently been advanced in the critical infrastructure debates, where the focus has shifted from protection to resilience. This development reflects the acknowledgement that complete protection can never be guaranteed, and that achieving the desired level of protection is normally not cost-effective in relation to the actual threats. Implicitly, this resilience discourse emphasises not prevention but the ability to recover from a failure or a disaster (e.g. Pursiainen and Gattinesi, 2014).

In spite of the scepticism, the safety culture approach has largely been adopted, and applied in many industries. In aviation, by way of illustration, a culture aimed at avoiding human error in particular is called Crew Resource Management or Cockpit Resource Management (CRM). This is a training system that aims at providing the best operational culture for the avoidance of human error. The main philosophy of this management culture is that instead of adopting a hierarchical system between the first pilot and other crew members, everyone is encouraged to take responsibility for safety and security factors. In the case of conflict in decision-making, effective conflict resolution is supposed to be focused on what is right rather than who is right. Thus, active participation in decision-making processes should be encouraged and practised, including questioning the actions and decisions of one's superior. Crew self-critique is another essential component of effective group processes, and it is emphasised that critique is not the same as criticism (e.g. Helmreich and Foushee, 2010).

3.2 Risk avoidance and removal

As discussed in the previous chapter, not all risks are selected for treatment – the risk might be considered too low, too improbable or too expensive to treat, for instance – but those which are, are supposed to be the subject of a treatment plan. The purpose of a risk treatment plan is, first, to justify the risk treatment options and combinations that are to be used. One should present the reasons for selecting the treatment options, including the expected benefits to be gained. Second, the plan has to document how the chosen treatment options will be implemented: who is accountable, what the resource requirements are, how the performance is measured, what the constraints are, the timing and schedule and similar issues. Monitoring the functioning of the plan needs to be an integral aspect in order to provide assurance that the measures will remain effective. A significant risk in and of itself can namely be the failure or ineffectiveness of the risk treatment measures. Risk treatment itself can also introduce new risks, either of the relatively simple direct variety or difficult-to-identify secondary, unintended risks that need to be assessed, treated, monitored and reviewed.

The ISO 31000 standard lists the options for risk treatment from a typological perspective: risk avoidance, risk removal, diminishing the likelihood of the risk, preventing or mitigating the consequences of the risk, risk sharing and retaining the risk. There is also the option of conscious risk-taking as a separate risk treatment alternative. This does not sit very well as a prevention and mitigation strategy, but is rather applicable as a profit-making strategy in a business environment, or in a situation where one has to take one risk in order to avoid another. Most risk treatment options are not exclusive, but they could be considered and applied either individually or in combination. The ISO 31000 assumes that an organisation normally benefits from the adoption of a combination of treatment options, as there is always uncertainty related to the treatment option's efficiency.

Choosing the most appropriate risk treatment option or options involves balancing the costs and efforts of implementation against the benefits derived. This might include considering legal, regulatory and other requirements such as social responsibility and the protection of the natural environment, as well as justifying the treatment option on economic grounds. When selecting risk treatment options, the organisation should consider its own values and perceptions as well as those of its stakeholders, be they related agencies, shareholders in a company, clients or citizens and the public at large. Certain risk treatment options may be more acceptable to some stakeholders than to others, and this acceptance might also be reflected in the efficiency of the treatment. Therefore, the information provided in treatment plans should include the reasons for selection of treatment options. The plan should also clearly designate those who are accountable for approving the plan and those responsible for implementing it, the concrete proposed actions, resource requirements, performance measures and constraints, reporting and monitoring requirements, as well as the timing and schedule. As such, the plan should then be integrated with the management processes of the organisation, including communication with appropriate stakeholders.

Risk-taking is regulated

Risk avoidance is the first and most obvious solution, if possible and practicable. It entails eliminating any exposure to risk that poses a potential loss, which is best applied by not performing any activity that may carry risk. In a financial company, for example, the risk assessment may have revealed that there is a high risk associated with the production of oil as well as a credit risk associated with the oil companies in the context of environmental risks and the success of renewable energy. An investor may consequently decide to avoid having a stake in the company, albeit taking a new risk at the same time by forfeiting any potential gains the oil stock may yield. Risk avoidance is thus closely related with risk perception, discussed in the previous chapter. In simple cases, the organisation should just decide, starting from its perceptions and values, whether the simplest approach would be to avoid the perceived risk altogether.

But risk avoidance is also very much related to regulations, and is not only a question of internal decision-making in an organisation. Almost any action may carry risks, not only for the respective organisation but also its clients or society at large. This in a way represents a move from the pure organisational risk management of, say, a private company or individual public agency, to that of risk governance involving a much broader spectrum of stakeholders, as briefly discussed in the previous chapter. In societal risks pertaining to safety and security, where high risk-taking is not usually an option, there are naturally regulations, legislation, building codes, standards, certifications, and so on to this effect, precisely designed for the purpose of avoiding risks. National and international regulations in public health, environment, construction, spatial planning, and so forth place limits on those risks that can be tolerated and those that should be completely avoided, often by proscription and prohibition. Even in a commercial-technological field such as Information and Communication Technology (ICT), risk avoidance implies "regulatory intervention" (Jachia and Nikonov, n.d., p. 3), as there might be no incentive otherwise for profit-seeking companies to avoid risks for their clients.

Regulation can also mean establishing "liability rules based on the notion that organisations should internalise the costs of the risks they produce and that by internalising them, they will make wiser choices about the technologies they use" (Egan, 2007, p. 14). This, in essence, would necessitate a well-functioning tort liability legislation, which would make it easy for consumers, both public and private, to subsequently demand compensation for losses created by risks that were not of their own making. This, in turn, would force both the public and private sector to pay more pre-emptive attention to risk-taking because of their self-interest. However, it has been argued that governments in most cases lack a coherent strategy to use regulatory tools in managing risks (OECD, 2010). On the other hand, it has also been pointed out that absolute safety from risks "cannot be a sensible regulatory goal" (Majone, 2010, p. 94).

The precautionary principle

The most famous – some might say infamous – risk avoidance strategy is the so-called precautionary principle, which merits a short discussion in this context. The concept emerged in European environmental policies in the late 1970s. Since then, it has been used in many United Nations conventions, remaining today a highly influential regulative principle also in many other contexts. While quantitative techniques could be used to support the application of the precautionary principle, in essence it is instead a matter for qualitative evaluation. It was in the early 1990s that precautionary strategies became "the new buzzwords of the environmental protection field" (Cameron and Abouchar, 1991). From an academic point of view, the issue was the subject of a lively discussion in the literature, especially at the beginning of the 2000s, due to its institutionalisation in many countries' legislation. In the European Union 2000 legislation, it was broadened and duly applied beyond environmental protection, especially in relation to possible dangers to human, animal or plant health (European Commission, 2000). According to this regulation, the precautionary principle may be invoked when a phenomenon, product or process could have a dangerous effect, identified by a scientific and objective evaluation, but this evaluation does not allow the risk to be determined with sufficient certainty.

Because of its disputed nature, there is no clear-cut definition of the precautionary principle, however. Instead, it is formulated in different ways by different interest groups. Thus, the so-called strong precautionary principle states that no action should be taken unless one is sure that it will do no harm, whereas the weak precautionary principle states that it is unjustified to abstain from taking action if one is uncertain whether it may or may not be harmful. Recalling that risk is the combination of likelihood and consequence, one can thus see that the strong precautionary principle attaches a great deal of importance to – potentially extremely harmful and possibly even irreversible – consequences rather than the likelihood of these consequences occurring, which may be difficult to establish with sufficient accuracy (Pearce, 1994, p. 132; Morris, 2000, pp. 106–107).

Ultimately, the precautionary principle is therefore aimed at uncertain risks, the uncertainty usually stemming from complexity (cf. Van Asselt and Vos, 2006, p. 314). Related to action or inaction in the face of uncertainty is the notion of the burden of proof, stating that the proponent of an action must also be the one to prove that no harmful consequences will ensue as a result. On the other hand, the burden of proof can also be shifted to the opponent of the action. This party will then have to demonstrate that an action must not be taken because of overwhelming and conclusive evidence as to its harmful consequences. In a way, this creates a dilemma in the conditions of uncertainty or risk. It has been noted that the concept of the burden of proof may be vague without considering the standard of proof, which can be based on an analysis of the balance of probabilities (Morris, 2000, p. 10; Cameron and Abouchar, 1991, p. 22).

Therefore, in order not to let this regulation remain arbitrary, the European Union (European Commission, 2000), for instance, has set some standards. The use of this principle should be informed by the fullest possible scientific evaluation and the determination of the degree of scientific uncertainty. Furthermore, there should be a risk evaluation of the respective action, and also an evaluation of the potential consequences of inaction. In addition to all of this, it is further demanded that all interested parties participate in the study of precautionary measures, once the results of the scientific evaluation and the risk evaluation are available. In the whole process, one has to consider the proportionality between the measures taken and the chosen level of protection, non-discrimination of any parties, as well as the consistency of the measures with similar measures already taken in comparable situations. Finally, in the light of new scientific developments, measures taken on the basis of the precautionary principle should be reviewed. The above-mentioned European Union regulation also includes some general guidelines about the burden of proof in case this is not regulated in the legislation.

The issue of the burden of proof reveals the contested nature of any single case of risk avoidance based on the precautionary principle. There are always winners and losers. For instance, it is claimed that the principle has been utilised subjectively to benefit, say, the European Union against American goods, and that it has "engendered endless controversy" (Foster, Vecchia and Repacholi, 2000). In this context, uncertainty is often perceived as something that can be eradicated or reduced by research, monitoring or the passage of time. This implicit or explicit starting point has, however, been challenged by the claim that more knowledge does not, by definition, mean less uncertainty and vice versa. Instead, the fundamental imperfection of knowledge is the essence of uncertainty. As Van Asselt and Vos (2006) put it, this has led to a paradoxical situation. While it is increasingly recognised that science cannot provide decisive evidence on uncertain risks, policy-makers and authorities increasingly resort to science for more certainty and conclusive evidence. Klinke and Renn (2002), in turn, argue that the decision to use the precautionary principle depends on the type of risk. Risks related to human interventions in ecosystems, technical innovations in biotechnology and the greenhouse effect, persistent ecosphere pollutants and endocrine disruptors are examples of risks that may justify the use of the precautionary principle.

Vehement criticism of the whole principle at least in its strong form has argued that the precautionary principle is altogether too precautionary. In many cases, it has become counterproductive to its own objectives as its strong form does not allow risks that may lead to better futures. Examples include arguments to the effect that it may slow down technological innovations that could actually be beneficial for the environment. Therefore, where the burden of proof is on those who create a potential risk, the principle should be rejected, the argument goes, "not because it leads in bad directions, but because it leads in no direction at all" (Sunstein, 2003, p. 1003). Critics of the precautionary principle have pointed out that it also overlooks the bigger picture. It has been argued that the principle could even be seen as a risk-seeking strategy, as applying it in one industrial sector could consume resources that cannot then be allocated to equal or more serious risks in other sectors (Nollkaemper, 1996). Even within

48 Prevention

one sector, it is argued, the precautionary principle overemphasises possible losses and ignores the benefits. For instance, in the case of a medical drug, which is likely to produce high-risk side effects, these losses are emphasised at the expense of its life-saving effect. Thus, the precautionary principle

> tends to focus the attention of regulators on some particular events and corresponding losses, rather than on the entire range of possibilities. As a consequence, regulators will base their determinations on worst cases, rather than on the weighted average (expected value) of all potential losses and benefits.
> (Majone, 2010, p. 106)

Thus, the principle can actually lead to bad decisions, resulting in over-protection, for instance. By putting extensive restrictions on child-adult or child-child relationships in the name of child safety, as one example goes, this excessive protection is actually hindering children's development in that they will be deprived of normal social or human relations (Guldberg, cited in Morris, 2000, pp. 127–139).

Those who defend the principle often refer to potentially irreversible damage, a risk that cannot be taken. The point is that there is insufficient scientific backing for a certain risk perception. Kriebel et al. (2001), however, point out that there is a complicated feedback relationship between scientific discoveries and the setting of policy. Scientists should be aware of the policy uses of their work and of their social responsibility. While it is admitted that the precautionary principle in many cases suffers from imprecisions (Sandin, 1999) or is vague and ill-defined (Adams, 2002), it has also been claimed that the principle is no more vague or ill-defined than other decision principles, and can be made precise through elaboration and practice. It can furthermore be developed by adding specification of the degree of scientific evidence required to trigger precaution. While it may in certain circumstances lead to increased risk-taking by focusing on individual risks, this can be adjusted by broadening the framing of the problems. If the precautionary principle is value-based, and in that sense not completely scientific in any objective meaning, then so are all decision rules (Sandin et al., 2002).

Remove the existing risk sources – sometimes!

In ISO 31000 parlance, 'risk removal' is a clearly separate option from that of risk avoidance. It refers to eliminating the source of an identified, already existing risk. Some scholars, however, prefer to use the concept of risk avoidance to cover risk removal as well (e.g. Pritchard, 2015, p. 49). Indeed, it is sometimes difficult to draw the line. In any case, risk removal can be understood as an option for eliminating an existing risky technology or practice, for instance, or for completely changing the approach to the work in question.

An example is removing the risk of passive smoking by prohibiting smoking in public places through legislation (Sargent, Shepard and Glantz, 2004), or the removal of asbestos from old buildings (Lemen, 2004). The decision by Germany to shut down the old generation of nuclear reactors (and re-examine the safety of all national nuclear power facilities) after the 2011 Fukushima disaster was a clear risk removal option, at least as far as the older reactors were concerned. As many other European countries came to quite the opposite conclusion, this example also shows how risk treatment options are often related to diverging risk tolerance levels, affected by the political situation, media reporting, trust in alternative technologies and history (see Wittneben, 2012).

Risk removal may sometimes be the right option, and at other times not. A good example of a situation when risk removal has not been considered an optimal solution concerns that of the approximately 40,000 tonnes of chemical munitions that were dumped into the Baltic Sea after the Second World War. It is estimated that these munitions contain some 15,000 tonnes of chemical warfare agents. The munitions pose considerable risks to fishermen and workers involved in offshore construction activities, such as offshore wind farms, sea cables, pipelines, as well as the marine environment. Yet it has been recommended by experts that it is better not to try to remove these munitions, since it would be an even more dangerous risk than letting them remain where they are. Therefore, the risk treatment strategy is to continue mapping the dump sites instead and to circumvent them when executing underwater operations (CHEMSEA, 2014; HELCOM, 2013).

The above discussion has shown that both risk avoidance and risk removal can be highly complicated risk treatment options, encompassing not only uncertainty but often largely different interests and risk perceptions. However, risk avoidance and risk removal may be inconvenient risk treatment strategies for other reasons as well. First, some or perhaps most risks are simply not wholly avoidable or removable through purposeful treatment. This concerns not only natural hazards and the like. As discussed in the previous chapter, many risks are not identifiable in the first place (Bellavita, 2006; Gundel, 2005; Klinke and Renn, 2002). Second, risk avoidance or risk removal may become too expensive and/or impractical. If one considers that a certain security risk would be avoidable simply by ramping up the protection level to the maximum, in the critical infrastructure protection field, for instance, a small increment in the level of protection might incur significant additional costs. The conclusion might be that it soon becomes meaningless to invest in "specific infrastructure upgrades to avoid certain risk scenarios which may or may not occur" (De Bruijne and Van Eeten, 2007).

3.3 Reducing the likelihood and consequences of a risk

Reducing the likelihood of risks and reducing the consequences of a risk are two different options in ISO parlance. However, they lend themselves to being discussed together and are often implemented in tandem. Although the subject should not be oversimplified, changing the likelihood is more about focusing on prevention, whereas changing the consequences is about mitigation. The so-called Bow-tie diagram was introduced in the previous chapter. The same diagram can be used to illustrate these two risk treatment strategies, as shown in Figure 3.1. The aim is to define the top event, and then add so-called

50 *Prevention*

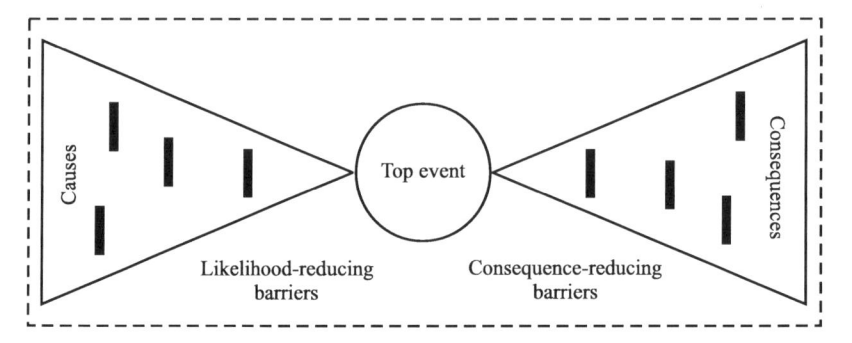

Figure 3.1 Bow-tie diagram with barriers

barriers (or, as often expressed, controls, safeguards or using tools) to both the causes side and the consequences side.

Like any risk treatment options, the current ones also assume that one has first conducted a proper risk assessment, which in turn presupposes that these risks are known to us, and not random risks emerging unexpectedly, and that one has evaluated them as requiring treatment. While even in these circumstances, changing the likelihood is usually not possible vis-à-vis natural hazards such as disastrous storms or tsunamis – "acts of God" as they were once called in insurance contracts – one can usually build barriers for the likelihood of cascading risks such as floods, for instance, by spatial planning or building walls, which in this case would then be the consequence-reducing barriers of the top event.

Yet the general crisis management literature does not provide much assistance with regard to understanding what exactly should be done in terms of reducing the likelihood and consequences. Rather typically, Haddow, Bullock and Coppola (2011, pp. 72ff.) propose "sustained action to reduce or eliminate the risks to people and property". The vagueness of these types of definitions is understandable because it always depends on the risk and on the context in which we discuss it. Barriers can therefore be of almost any description. Drennan et al. (2015, p. 108), focusing on public sector risk management, argue that the types of "treatment available to reduce risks include physical measures/barriers, changes to management systems, human resources strategies and the use of contracts". Haddow, Bullock and Coppola (2011, p. 75), focusing on emergency management, state that "the following mitigation tools are known to reduce risk: Hazard identification and mapping; Design and construction applications; Land-use planning; Financial incentives; Insurance; Structural controls."

In many cases, in order to identify the barriers that need putting in place, one has to think about the opposite, namely, the vulnerabilities. Usually these vulnerabilities have already been identified during the risk assessment phase. Like the barriers, the vulnerabilities may take almost any form, and are more likely to be a combination of many vulnerabilities that may materialise when

Table 3.1 Early warning vulnerability factors

Early warning vulnerability factors (examples)	Sub-factor (examples)	Likelihood reducing	Consequence reducing	In the realm of the organisation's influence
Human factors	Human-machine interaction	Yes	No	Yes
Technological factors	Early-warning detectors	No	Yes	Yes
Organisational factors	Preventative maintenance system	Yes	No	Yes
Public policy, societal and cultural factors	Regulative risk reduction	Yes	Yes	Often not

events coincide. One way to generalise this discussion is to differentiate between vulnerability factors at a more generic level. This might then work as a kind of checklist to see what kind of barriers need constructing or enhancing. A blueprint of this type of checklist, coupled with some illustrations, is presented in Table 3.1. This kind of systematic consideration of vulnerability factors, usually emerging when preparing the risk assessment and respective barriers, could be tailored by the risk manager according to needs. If the perspective is that of one organisation aiming at building risk barriers, it might be rational to differentiate between the barriers that the current organisation can influence and those that are largely beyond its control.

Let us briefly discuss some of these vulnerability factors and the specific issues mentioned in Table 3.1 and consider barriers that could be put in place to prevent these vulnerabilities from materialising.

Human factors

Imagine that, in principle, the technological, organisational and institutional, economic, political and other similar circumstances would allow for the prevention or mitigation of a crisis. Nonetheless, the crisis or disaster occurs. In many cases, one then finds a reason to apportion blame to the so-called human factor. The issue concerns human performance, affected by many attributes such as age, physical health, state of mind, attitude, emotions, and so forth. If one omits intentional human actions, such as terrorism, sabotage or deliberate negligence, the focus will be on the limits of a human being's capacities, particularly related to information processing.

A piece of the puzzle is so-called human error. The fact that human errors have contributed to some of the most serious infrastructure and technological disasters, such as Three Mile Island in 1979 and Chernobyl in 1986, has been suggested in the research literature for quite some time (Senders and Moray, 1991; Reason, 1990), and continues to be a major theme (e.g. Saeed, Bajwa and Bakwa, 2014). This leads to the so-called situation awareness debate, which focuses particularly on ergonomics, especially on a human being's limited capacity to handle information when functioning in a highly developed technological environment. A widely accepted definition of (successful) situation awareness includes three levels: obtaining data from the environment; understanding what this data means, based on knowledge or experience; and finally being able to make predictions about what will happen (Endsley, 1988, 1993, 1995, 2013; Salas et al., 1995). However, most of the literature on situation awareness is actually about situation *un*awareness, about the human failure to understand a situation correctly, which then leads to poor or ill-informed decision-making and behaviour (e.g. Bolstad, Costello and Endsley, 2006).

One of the most discussed phenomena of this kind is so-called attentional (or cognitive) tunnelling. In a potential early warning situation, the number of data sources can be high and their relative importance may change. If one then, for one reason or another, does not allocate one's attention between the channels of information, or diagnostic hypotheses, or task goals, in a maximal way, the result may be a fixation on one specific element while becoming blind to the other elements (Wickens, 2005). Another phenomenon is a so-called requisite memory trap, referring to the limited working memory capacity of a human being, concerning how much short-term information such as numbers, variables, and so on a human being can hold in their mind simultaneously. Although people differ in this respect, and working memory can be developed by training, every human being has their limits in this sense. Sometimes this is not understood properly by technological system developers from the perspective of exceptional situations. While the system can work perfectly in normal conditions, it occupies all of the working memory of its operator, which may mean that he does not notice possible signals intimating a crisis situation before it is too late. A slightly different type of human-technology problem is that of data overload, or rather its overly complex presentation. For example, a modern civil passenger aircraft has approximately 450 different types of warnings (Stanton and Edworthy 1999, p. 348). If a system, such as a dashboard for potentially dangerous machinery, allows for too many variables and simultaneous functions, it can lead to failures in information processing in both normal and potential crisis situations. The opposite case to data overload may be that of the so-called out-of-the-loop syndrome, where a highly automated system removes the operator too far from the elements he controls, to the point where he loses touch with them (Moulton and Forrest, 2005).

Thus, enhancing the technological systems' ergonomics from the perspective of the human capacity to process information, including crisis situations, is necessary. It is neither easy nor unambiguous, however, to design or choose the best system. For instance, while in general, auditory warnings are superior to visual warnings in reducing response time, in some cases they fail; they may be hard to learn or remember, or they may be hard to localise, and so on

(Stanton and Edworthy, 1999). Prolonged testing might be needed in trying to locate the possible human factor vulnerabilities in high-risk systems, often also called high-reliability systems for that very reason (cf. Perrow, 1984; La Porte and Consolini, 1998). Therefore, in this case, the likelihood-reducing barrier would involve designing technological machinery in a human-error–reducing way, including enhanced testing from this particular perspective.

The other side of the coin is to concentrate on the very system operators. Human errors can be prevented or reduced by such barriers as proper training, procedures, memory aids, intelligent decision-support systems, self-knowledge of one's error potential – in short, different kinds of error management elements (Reason, 1990, pp. 234–257). As early as 1988, Weick, in turn, drew attention to the concept of sensemaking, meaning that better understanding of how human action itself can create a crisis or at least contribute to its worsening after the triggering event could be helpful for organisations in preventing negative consequences.

Technological factors

Science and technology play a central role in forming the basis for risk assessment and prevention. Let us consider technological early-warning solutions, such as automatic sensors, detectors and surveillance systems. General technological developments have given rise to equipment-based technological solutions for monitoring, detecting and analysing almost anything, be it radiation; drinking water quality, air quality or specific substances in the air; water level; smoke or heat; air, road and sea traffic; storms and hurricanes, earthquakes and tsunamis; oil spills; health-related threats such as pandemics; financial risk indicators, and so forth. There are several specific remote sensing and geographic localisation and information systems available to define exactly where a given emergency is taking place and who should receive an early warning. Satellite and other wireless communications systems help to transmit early warning information speedily between authorities and to the people at risk, thus providing more room for prevention, warning and early response. There are automatic damage mitigation and response systems, which turn off dangerous functions, turn on emergency lighting and exit signs, create loud audible warnings, turn on sprinkler systems, unlock emergency storage facilities and equipment cabinets, start automated control of lifts and air-conditioning systems, and so on. An increasing number of new high-tech solutions, such as using mobile SMS messages as early warning tools, or using drones in natural catastrophe detection, are constantly being introduced (e.g. Ghosh, 2012).

Let us illustrate the use of technology as a likelihood- or consequence-reducing barrier by means of a drinking water system. While in developed countries there currently exist organisational structures and scientific methods to provide a high-level control mechanism for environmental water bodies and drinking water quality, these methods are designed for long-term decision-making and generally not for an immediate response in case of an incident. Imagine now that our risk assessment has, however, considered that a terrorist attack on our drinking water system using, say, anthrax or other dangerous chemical substances, is a real risk. One therefore has to build barriers against that risk. While one may not be able to prevent the attack, technology provides tools for early warning, in that one may be able to reduce the consequences of the attack. Innovative water quality monitoring systems – automatic sensors and detectors – have been developed in the last few years, which allow for real-time control of the overall water quality. These systems react to a number of classes of contaminants and immediately warn operators of potentially malicious or non-malicious contamination in the network (Tanchou, 2014; Hufnagl, 2013; Raich, 2013; Llorca and Rodríguez-Mozaz, 2013). This kind of detector system might completely prevent the intended consequences or at least mitigate the number of affected areas and people.

Organisational factors

It goes without saying that organisational matters can either cause or allow, or alternatively, prevent and mitigate, unwanted events and crises. For instance, it is widely held that the September 11, 2001 terror attacks could have been prevented if communication within and between the responsible organisations had worked, instead of their being reluctant to share information with each other (Parker and Stern, 2002, p. 611). While there are a countless number of organisational factors, let us focus on one which is highly concrete and usually present in most technology-related organisations, namely, maintenance. This concept is related to technological systems, with the question being how to organise it. Resistance to disturbance can be considerably enhanced by introducing a good maintenance system as part of organisational routine. According to the European Standard EN13306:2001 (EN, 2001), maintenance is defined as follows: "A combination of all technical, administrative and managerial actions, including supervision actions, during the life cycle of an item intended to retain it in, or restore it to, a state in which it can perform the required function." This implies that maintenance is also strongly related to the restoration of a system or an item, not only preventing it from failing.

There are numerous ways to categorise maintenance, but the simplest is to say that it can be either preventive or corrective (Barabady and Kumar, 2008). While preventive maintenance means that the maintenance is performed in advance at set intervals to prevent the failure from occurring, corrective maintenance means that the components are run until they fail (Moubray, 1997). Should one consider preventive maintenance, it is usually planned, but also includes unplanned maintenance activities. There are a number of different maintenance strategies and techniques in existence, such as Reliability Centred Maintenance (RCM), Condition Based Maintenance (CBM) and Total Performance Maintenance (TPM) (Garg and Deshmukh, 2006).

When using CMB, for instance, the repair and replacement of components are based on the actual or future condition of the asset (Raheja et al., 2006).

The maintenance intervals are determined based on the reliability and criticality of the component. This approach is supposed to lead to an optimal balance between cost and maintenance frequency. In order to acquire the best knowledge about the current condition of the asset's components, inspection can often be carried out as a useful tool. An infrastructure typically has a high number of components that should be subject to inspection. Considering the limited amount of resources, one should therefore establish a priority list of the components that should be inspected. In order to do this, a decision-making technique called Risk Based Inspection (RBI) can be used as an additional helpful tool (Moura et al., 2015; DNV, 2010; Khan, Haddara and Bhattacharya, 2006). This approach proposes that planned corrective maintenance actions will be issued for low-risk systems and components, whereas systems and components with a high risk (or mere criticality of consequences) will be issued for preventive maintenance tasks based on condition monitoring.

Thus, in our illustration, the vulnerability would be the failure of a technological system, whereas our preventive barrier would be a well-thought-out preventive maintenance strategy and implementation system. In order to develop the performance of the maintenance processes, it is desirable to establish a well-formulated maintenance strategy based on a corporate and manufacturing strategy (Muchiri et al., 2011), using strategies such as CBM and RBI. The main objective is to define a unified maintenance process within the organisation, which can be regarded as a critical success factor for achieving manufacturing and business success. The main objective of planned maintenance is to ensure that the condition of the asset is as good as possible considering the available resources. In that way, when an unwanted event occurs, such as flooding, or a terror attack, for instance, the infrastructure will be less affected, and the recovery process will also be more efficient.

The challenges of regulative risk reduction

Let us consider just one illustrative field from the perspective of regulative risk reduction, namely, critical infrastructure protection and resilience in respect of those infrastructures that are necessary to maintain vital societal functions such as electricity, transport, health care and supply chains (cf. Pursiainen, 2009). Regulation refers in this context to the legal obligations set by states for the owners and operators of critical infrastructure to prevent any disturbances or to mitigate their consequences. Governments are usually legally responsible for safeguarding critical infrastructure, although most of the latter is owned, administered and operated by the private sector. This is why public–private partnership (PPP) is considered a major issue in safeguarding national infrastructure (e.g. Abele-Wigert, 2006, pp. 57–58). While in the United States private industry traditionally owns most of what is defined as national infrastructure, its share being estimated at 85 per cent, in many European countries such infrastructure as water, energy and railway transportation has previously often been taken care of exclusively by the government. However, since the 1980s there has been an ongoing process of market liberalisation and privatisation of these infrastructures. The rapid development of predominantly privately owned and operated ICT, and the dependence of other sectors upon it, have complicated the situation. This has led to a rather ambiguous situation in terms of the real authority. While government authorities

> may have, formally or informally, the overall responsibility for the reliable provision of services, they lack the authority and resources to actually fulfil that responsibility. Central governments bodies and policy makers involved [in critical infrastructure protection] to a large extent lack the technical expertise and the means to monitor or control CI [critical infrastructure] operations.
> (De Bruijne and Van Eeten, 2007, p. 24)

Therefore, private industry is supposed to be able to exert extensive self-regulation, because only they in practice have access to the necessary technical capabilities and information for the majority of critical infrastructure protection and resilience.

Globalisation, with its tendency to move private companies outside the nation state, has moreover made the situation more complex from the perspective of government control. The fact that national critical infrastructures are dependent not only on other sectors but on the situation of other countries' critical infrastructures complicates the situation, because no single country is either immune to the effects or able to predict the outcomes if neighbours suffer from serious infrastructure disruptions (cf. Mussington, 2002, pp. 25, 26).

Here we face the dilemma of the common good. Indeed, De Bruijne and Van Eeten (2007, p. 24) have noted that while PPP may seem self-evident and is celebrated by all parties, this "shallow consensus" is usually broken when it becomes clear that governments expect the private sector to make considerable investments beyond their cost-benefit calculations. They claim that this dilemma leaves governments with only two options: to provide the necessary resources, funded from the public budget, or to increase regulation. According to De Bruijne and Van Eeten, the first option is largely impossible, mainly for financial resource reasons but also for reasons related to the necessary separation of public funding from private rent-seeking use. The second option, adding regulation, would force the private sector to allocate more resources to deal with the protection or resilience of the systems they own or operate. Egan, among others, has proposed that, because markets are at present externalising the critical infrastructure risks, state regulation should mean establishing "liability rules based on the notion that organizations should internalize the costs of the risks they produce and that by internalizing them, they will make wiser choices about the technologies they use" (Egan, 2007, p. 14). This, in essence, would necessitate a well-functioning tort liability legislation, which would make it easy for consumers, both public and private, to subsequently demand compensation for losses created by critical infrastructure failures, which in turn would force industry to pay more pre-emptive attention to security and protection because of their self-interest.

However, De Bruijne and Van Eeten (2007, p. 25) claim that increasing state regulation would mean coming "full circle" (from liberalisation back to state regulation). Indeed, while there might be some willingness in Europe to increase regulation in respect of critical infrastructure, in the United States this approach has traditionally been seen as unwise, presumably based on the ideological commitment to the idea that the state's role in business should be minor. Thus, the Presidential Decision Directive concluded back in 1998, for instance, that while the PPP "must be genuine, mutual and cooperative", it means that "the US government should, to the extent feasible, seek to avoid outcomes that increase government regulation or expand unfunded government mandates to the private sector" (PDD, 1998, p. 2). Indeed, as De Bruijne and Van Eeten (2007, p. 25) argue, when governments have the two options – that of providing the necessary protective and resilience resources themselves, or that of adding state regulation – most strategies propose neither. Instead, national strategies are usually confined to the status quo by advocating mere awareness-raising, best practice exchange and soft "commitment power" efforts with regard to private actors.

In a way, the PPP is a typical dilemma in that it is worse than a mere problem, as there are no good solutions available. In these conditions, government involvement in practical efforts to secure the critical assets in the private sector remains rather limited. As Robinson, Woodard and Varnado (1998) have noted, the natural starting point is that private industry determines investments in protecting infrastructure from a business perspective. However, at the same time as the vulnerabilities are increasing, it has been recognised that while market liberalisation supports policies which emphasise the importance of low prices for consumers, one 'side effect' has been to reduce the funds available for investment in and maintenance of key assets (IRGC, 2005, pp. 1–2). Thus, neither private industry's self-regulation nor market forces seem to provide the necessary level of protection. Indeed, security has never been a design driver for market forces in their dealings with critical infrastructure (Dunn, 2006, pp. 29–30). However, Robinson, Woodard and Varnado (1998) propose that in order to have a proper security strategy, it is important for industry to have all the information it needs to perform a risk assessment. The primary focus of industry-government cooperation should therefore be to share information and techniques related to risk management assessment, the identification of weak spots, plans and technology to prevent attacks and disruptions, and plans for how to recover from them.

Hurley (2000, p. 4) argues that within this cooperation the private sector's role would be to help government authorities in risk analysis and technical issues, thereby assisting them to arrive at practical, workable solutions to real challenges more quickly. Mussington (2002, p. 31), in turn, argues that this kind of information sharing should be non-hierarchical. A decentralised, confederated response and information-sharing mechanism for enhancing information assurance seems to provide a more flexible means of meeting a fast-changing threat to infrastructure vulnerability than top-down methods of managing information assurance. However, Robinson, Woodard and Varnado (1998) have further pointed out that in practice there are some obstacles to this information sharing from the private sector side to that of the government authorities. From the private company's point of view, collaboration may include or require passing over classified and secret material, proprietary and competitively sensitive information, liability concerns, fear of regulation and legal restrictions. These issues make far-reaching PPP difficult in practice.

On the other hand, Hurley (2000, p. 4) sees a possibility here. The active participation of the private sector in the development of critical infrastructure protection and resilience strategies would help in promoting general acceptance by the private sector of any regulatory approach that government may find necessary to adopt. In practice, according to Hurley, this PPP in the development of strategies can occur in at least the following ways: providing comments on proposed government regulations; participating in the work of advisory committees to government agencies; serving in voluntary groups that research and draft publications germane to ICT issues; and participating as speakers or panelists in forums.

The European Union, for instance, has adopted the above-mentioned compromise, thus avoiding any far-reaching regulation of the private sector. The Directive of the European Council from 2008 stresses that effective protection requires communication, coordination and cooperation nationally and at the European Union level, involving all relevant stakeholders. However, while the chosen approach is supposed to "encourage full private sector involvement" (European Council, 2008, p. L345/76), in practice, however, the directive only demands that each owner or operator of European critical infrastructure should establish an Operator Security Plan identifying critical assets and proposing relevant security solutions for their protection. Moreover, a Security Liaison Officer should be appointed and the Operator Security Plans should be forwarded to the ECIP Contact Point in the Member States.

Thus, this seems to confirm Andersson and Malm's (2006, pp. 166–167) argument that PPP in its current form as structural cooperation between equal parties is seen by both public and private actors as the most effective way to reach their goals. Fioritto and Simoncini (2011, p. 130) formulate this into a normative rule to the effect that "the development of market-based incentives that can match the traditional administrative enforcement by stimulating private parties comply with rules by themselves". For governments, PPP provides a means of engaging the private sector in public affairs and achieving guidelines and standards without having to use strict regulatory means. For private actors, PPP offers a flexible way of meeting government requirements while avoiding regulation.

3.4 Risk sharing

Sometimes a risk might be difficult to avoid or remove. Its causes may also be difficult to limit and its consequences largely uncontrollable, involving particularly high risks. There is, however, a positive risk treatment option for these cases: the risk can be shared! While the risk with all its consequences remains, by sharing or diversifying, one is able to limit the consequences for an individual actor or organisation. In a way, this option simultaneously allows a certain amount of risk taking in circumstances of uncertainty, with the possibility of both losses and gains.

Risk sharing is the mainstream in businesses, but also in public and household activities. This may refer to a joint venture based on risk-sharing agreements, or risk pooling between stakeholders, but also to an insurance policy. In corporations, a well-developed insurance policy is an essential part of risk management (see e.g. Harrington and Niehaus, 2004). In principle, insurance means that the financial uncertainty of a potential future loss is replaced by the certainty of insurance cover, in return for an agreed premium. The premium is usually calculated based on historical data. The insurance company may also demand some risk controls or barriers to be in place in return for the policy or a lower premium rate. However, some uncertainties usually remain, to the effect that the risk perceptions and interests between the partners may diverge, creating conflicts or under- or overinsurance (Drennan, McConnell and Stark, 2015, pp. 111–113; Haddow, Bullock and Coppola, 2011, pp. 82–85). Also, there is no insurance market to cover all the possible risks that an organisation might have identified.

Yet, a developed risk-sharing system in general, at the level of society, can be understood as a positive feature, which is often a precondition for economic growth and social security. The potential loss perspective of risk is linked to the advancement of insurance practices where the central idea is to diversify significant exposures across a larger number of individuals, households and business entities. The current practice is extremely complex, whereby the global reinsurance market allows for a broader distribution of exposures among players in different parts of the world so regional disaster exposures can be shared. Obviously, this brings new risks with it as well, which are not easy to foresee.

3.5 Risk retaining

One risk treatment option is to do nothing. In ISO parlance, this is called "retaining risk by informed decision" (ISO, 2009a, p. 18). This is not a theoretical option, but is intimately connected to risk evaluation, already discussed above, noting that issues such as individual and group risk tolerance and perceptions play a great role. Still, the issue remains of how one in practice should draw the line for this risk tolerance. Often risk evaluation is justified in terms of principles rather than techniques, although, in part, one speaks about the same issues with different labels. In any case, risk retaining principles include: Appropriate Level of Protection (ALOP), Best Available Technology (BAT), Reasonable Relationship (balancing non-monetary benefits and the monetary costs of achieving them), Tolerable Level of Risk (TLR), and As Low As Reasonably Practicable (ALARP) (Yoe, 2012, pp. 76ff).

Let us consider perhaps the most well-known of these, namely, the ALARP principle. The starting point is that one has already conducted the proper risk

60 *Prevention*

analysis but then considers that certain risks should not be treated, usually because the risk is so low either in terms of its consequences, its likelihood, or both, so that no further risk reduction is considered necessary. High risks, on the other hand, should always be treated if possible. However, in between the low and high risks there is the middle ground, where one has to decide whether the risks should be treated or tolerated.

ALARP is a maxim to this effect. The idea is to reduce risks to a level that is as low as reasonably practicable, hence its name. In order to become subject to application of the ALARP principle, risks should be located in the ALARP area or region; a risk becomes relevant only if it is below the maximum tolerable level but above the low risk level. This is often illustrated in terms of an inverted pyramid where the upper part consists of intolerable risks, the lower part of acceptable risks and the middle ground of ALARP risks, as shown in Figure 3.2.

Sometimes ALARP is also expressed as the balance point or trade-off between risk mitigations and risk exposures that produces a point of equilibrium. ALARP is the level of risk that is tolerable and cannot be reduced further without expenditure or costs disproportionate to the benefit gained, or where the solution is impractical to implement (Talbott, n.d.). In other words, attempting to reduce a risk would be more costly than any cost likely to stem from the risk itself. In theory, then, it should be possible to find this equilibrium point, which would be the same as the intersection between the curve illustrating the risk level and the curve illustrating the costs (in terms of money, time, trouble, etc.) resulting from risk treatment.

High risk — Risk cannot be justified and has to be treated

ALARP-area — Tolerable only if risk reduction is impracticable or the cost of it would be highly disproportionate to the improvement gained

Tolerable if the cost of reduction would exceed the improvement gained

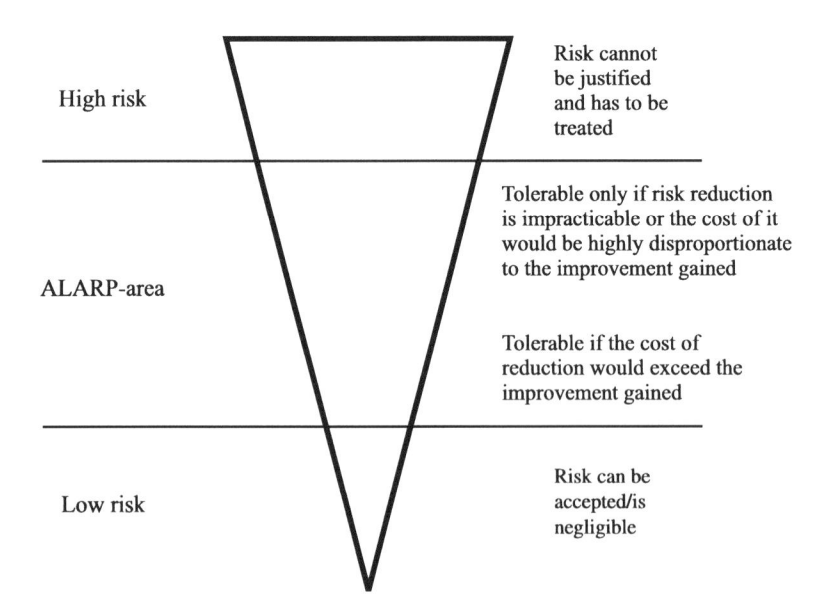

Low risk — Risk can be accepted/is negligible

Figure 3.2 The ALARP principle

ALARP has been criticised for being difficult to apply in practice. Demonstrating that all credible risk reduction methods are impracticable is often beyond the capacity of any single individual and therefore one always has to involve a larger group to apply ALARP. The second difficulty is that applying the ALARP principle fully and accurately easily becomes expensive even before one has started to treat the risks, that is, to identify those risks that are ALARP. In practice, the ALARP principle requires significant interpretation and can be very difficult and costly to apply (Hughes, n.d.). Others have argued that in striving to make specific industrial risks ALARP, for instance, one sometimes increases other risks; therefore, ALARP should be supplemented with a methodology that also considers whether or not there is a net increase or decrease in risks (Kletz, 2005). Instead of trying to make a single risk ALARP, the principle should be applied at the level of the aggregated or cumulative total risk (Wilkinson and David, 2009).

It has also been pointed out that ALARP, at least in safety issues, is often applied not in terms of a cost–benefit analysis, which would prescribe that a safety improvement should be undertaken only if the cost of doing so was less than or equal to the resultant benefits. Instead, ALARP would also allow improvements, the costs of which are not in gross disproportion to the benefits. This would include cases in which costs might substantially exceed benefits (Jones-Lee and Aven, 2011; cf. Melchers, 2001).

Overall, the main concern in this criticism is that, in fact, ALARP does not solve the issue of risk perception, providing a clear-cut objective mathematical solution for risk evaluation, but remains open to interpretations. However, there are some additional formal methodologies that help to take the divergent risk perceptions into account (French, Bedford and Atherton, 2005). Yet they might considerably complicate the practice of organisational risk assessment, and might therefore become too complicated and academic to be applied in normal organisational risk assessment practices.

Perhaps the main benefit of this approach is its very flexibility, however, although it concurrently places a great deal of responsibility on those drawing the limits and computing the costs versus sacrifices made to achieve a reduced risk level. To partially deal with this challenge, it is often helpful to rely on existing codes and standards, or good practices. On the other hand, existing regulations may sometimes be in conflict with the ALARP principle, demanding risk reduction over the limit where it is no longer reasonably practical to follow the ALARP principle.

3.6 Risk-taking as prevention?

Risk-taking is the opposite of risk avoidance, meaning that an activity that may give rise to a risk is consciously started or continued, or even increased, because the risk is not only seen as a threat, but also an opportunity (Newsome, 2015, p. 101). This risk option follows from the generic approach of the ISO standard, sometimes criticised by some safety and security-focused organisations (see Purdy, 2010, pp. 882, 885). The reason why risk-taking is understood as an option in risk-treatment plans in the ISO 31000 context is that the standard also considers such organisations as companies and institutions dealing with financial and commercial risks, which may often regard risk-seeking and risk-taking as quite normal and legitimate options (Coleman, 2011, p. 12). Indeed, when it comes to businesses, uncertainty is often regarded as an essential source of entrepreneurial value creation. However, it is often pointed out that successful risk-taking necessitates a corporate culture that facilitates and nurtures an urge to change and adapt organisational processes and respond when significant environmental changes are observed. Andersen, Garvey and Roggi (2014, pp. 151–152, 161) argue that consistently successful risk-taking "cannot happen by accident" and success in risk-taking "is as much a result of design as of luck". Moreover, firms that are successful at risk-taking "are also masters of the risk monitoring and strategic responsiveness processes".

Apart from financial risk-taking, individual risk-taking and, perhaps, risk-taking patterns in domestic and international politics, the literature on risk-taking is at first glance painfully sparse when it comes to risk and crisis management issues in, say, safety and security. While this option is not automatically excluded from such fields as civil protection, health care or operating high-risk industries, one must often take another risk in order to avoid the first one. The so-called risk society discourse and risk governance discourse, both briefly discussed in the previous chapter, operate on a more abstract level. In the traditional risk-society discourse it is all about the fact that the modern technological society has risks embedded in it, and is characterised by manufactured uncertainty (Beck, 1992). The somewhat newer risk society discourse considers risks as mostly constructed in society, where particular risks are expressed, legitimised, mediated or ignored (Adam, Beck and van Loon, 2000). The message is therefore not that one can necessarily choose or choose not to take (or avoid) a certain risk, as many risks are virtually embedded in modern society.

The more recent risk governance approach borrows to some extent from the risk society discourse, but is perhaps more practical in partially adopting the risk management terminology while being broader in terms of not looking at organisations but at society at large. This supports the notion that risk governance "requires risk(s)-benefit(s) evaluations and risk-risk trade-offs" (Van Asselt and Renn, 2011, p. 442). This statement follows from the idea that risks should not be considered in isolation, but there is always a more complex, uncertain and ambiguous holistic perspective to consider. This means that risk-taking is always an option because other risk treatment options – from a broader, holistic viewpoint – may produce a worse risk governance net result.

References

Abele-Wigert, I. (2006) Challenges Governments Face in the Field of Critical Information Infrastructure Protection (CIIP): Stakeholders and Perspective. In Dunn, M.

and Mauer, V. (eds) *International CIIP Handbook 2006. Vol. II. Analyzing Issues, Challenges, and Prospects*. Zürich: Swiss Federal Institute of Technology Zürich.

Adam, B., Beck, U. and van Loon, J. (eds) (2000) *Risk Society and Beyond: Critical Issues for Social Theory*. London: Sage.

Adams, M.D. (2002) The Precautionary Principle and the Rhetoric Behind It. *Journal of Risk Research*, 5(4), pp. 301–316.

Andersen, T.J., Garvey, M. and Roggi, O. (2014) *Managing Risk and Opportunity: The Governance of Strategic Risk-Taking*. Oxford: Oxford University Press.

Andersson, J. and Malm, A. (2006) Public-Private Partnerships and the Challenge of Critical Infrastructure Protection. In Dunn, M. and Mauer, V. (eds) *International CIIP Handbook 2006. Vol. II. Analyzing Issues, Challenges, and Prospects*. Zürich: Swiss Federal Institute of Technology Zürich.

Barabady, J. and Kumar, R. (2008) Reliability Characteristics Based Maintenance Scheduling: A Case Study of a Crushing Plant. *Reliability Engineering and System Safety*, 93(4), pp. 647–653.

Beck, U. (1992) *Risk Society: Towards a New Modernity*. London: Sage Publications.

Bellavita, C. (2006) Changing Homeland Security: Shape Patterns, Not Programs. Homeland Security Affairs, 2(3), pp. 1–22. Available at: http://www.hsaj.org/pages/volume2/issue3/pdfs/2.3.5.pdf

Blazsin, H. and Guldenmund, F. (2015) The Social Construction of Safety: Comparing Three Realities. *Safety Science*, 71, pp. 16–27.

Bolstad, C.A., Costello, A.M. and Endsley, M.R. (2006) Bad Situation Awareness Designs: What Went Wrong and Why? In *Proceedings of the International Ergonomics Association 16th World Congress*. Maastricht, the Netherlands.

Cameron, J. and Abouchar, J. (1991) The Precautionary Principle: A Fundamental Principle of Law and Policy for the Protection of the Global Environment. 14 B. C. *International and Comparative Law Review.*, 1, pp. 1–27.

CHEMSEA. (2014) CHEMSEA findings. Results from the CHEMSEA Project – Chemical Munitions Search and Assessment. Available at: http://www.chemsea.eu/admin/uploaded/CHEMSEA%20Findings.pdf

Choudhry, R. M., Fang, D. and Mohamed, S. (2007) The Nature of Safety Culture: A Survey of the State-of-the-Art. *Safety Science*, 45(10), pp. 993–1012.

Coleman, T. S. (2011) *A Practical Guide to Risk Management*. Middletown, DE: CFA Institute.

Cox, S. and Cox, T. (1991) The Structure of Employee Attitudes to Safety: A European Example. *Work and Stress* 5(2), pp. 93–106.

De Bruijne, M. and VanEeten, M. (2007) Systems That Should Have Failed: Critical Infrastructure Protection in an Institutionally Fragmented Environment. *Journal of Contingencies and Crisis Management*, 15(1), pp. 8–29.

DNV. (2010) *Det Norske Veritas, Recommended Practice DNV-RP-G101: Risk Based Inspection of Offshore Topsides Static Mechanical Equipment*. Baerum: Det Norske Veritas.

Dorothy, A. et al. (2009) Methodology for Assessing the Resilience of Networked Infrastructure. *IEEE Systems Journal*, 3(2), pp. 174–180.

Drennan, L., McConnell, A. and Stark, A. (2015) *Risk and Crisis Management in the Public Sector*. Second edn. New York: Routledge.

Dunn, M. (2006) Understanding Critical Information Infrastructure: An Elusive Quest. In Dunn, M. and Mauer, V. (eds) *International CIIP Handbook 2006.*

Vol. II. Analyzing Issues, Challenges, and Prospects. Zürich: Swiss Federal Institute of Technology Zürich, pp. 27–40.

Egan, M. (2007) Anticipating Future Vulnerability: Defining Characteristics of Increasingly Critical Infrastructure-like Systems. *Contingencies and Crisis Management*, 15(1), pp. 4–17.

EN. (2001) European Standard, EN 13306:2001, Maintenance – Maintenance Terminology.

Endsley, M.R. (1988) Design and Evaluation for Situation Awareness Enhancement. Human Factors Society. 32nd Annual Meeting 1988, Human Factors Society, Santa Monica, CA.

Endsley, M.R. (1993) A Survey of Situation Awareness Requirements in Air-to-Air Combat Fighters. *International Journal of Aviation Psychology*, 3(2), pp. 157–168.

Endsley, M.R. (1995) Toward a Theory of Situation Awareness in Dynamic Systems. *Human Factors*, 37(1), pp. 32–64.

Endsley, M.R. (2013) Situation Awareness. In Lee, J.D. and Kirlik, A.. *The Oxford Handbook of Cognitive Engineering*. Oxford Handbooks Online. Available at: http://www.oxfordhandbooks.com/view/10.1093/oxfordhb/9780199757183.001.0001/oxfordhb-9780199757183-e-5

European Commission. (2000) Communication from the Commission on the Precautionary Principle. Brussels, 2.2.2000 COM(2000) 1 final.

European Council. (2008) Council Directive 2008/114/EC of 8 December 2008 on the Identification and Designation of European Critical Infrastructures and the Assessment of the Need to Improve Their Protection. *Official Journal of the European Union*, pp. 345/75–82.

Executive Order. (2013) Executive Order 12636 – Improving Critical Infrastructure Cybersecurity. *Federal Register*, The POTUS of America.

Fioritto, A. and Simoncini, M. (2011) If and When: Towards Standard-based Regulation in the Reduction of Catastrophic Risks. In Alemanno, A. (ed.) *Governing Disasters: The Challenges of Emergency Regulation*. Cheltenham, UK: Edward Elgar, pp. 115–136.

Foster, K. R., Vecchia, P. and Repacholi, M. H. (2000) Science and the Precautionary Principle. *Science*, 288, pp. 979–981.

French, S., Bedford, T. and Atherton, E. (2005) Supporting ALARP Decision Making by Cost Benefit Analysis and Multiattribute Utility Theory. *Journal of Risk Research*, 8(3), pp. 207–223.

Garg, A. and Deshmukh, S.G. (2006) Maintenance Management: Literature Review and Directions. *Journal of Quality in Maintenance Engineering*, 12(3), pp. 205–238.

Ghosh, S. (2012) *Natural Disaster Management: New Technologies and Opportunities*. India: The Icfai University Press.

Guldenmund, F. W. (2000) The Nature of Safety Culture: A Review of Theory and Research. *Safety Science*, 34, pp. 215–257.

Guldenmund, F. W. (2010) (Mis)understanding Safety Culture and Its Relationship to Safety Management. *Risk Analysis*, 30(10), pp. 1466–1480.

Gundel, S. (2005) Towards a New Typology of Crises. *Journal of Contingencies and Crisis Management*, 13(3), pp. 106–115.

Haddow, G. D., Bullock, J. A. and Coppola, D. P. (2011) *Introduction to Emergency Management*. Fifth edn. Amsterdam: Elsevier.

Harrington, S. and Niehaus, G. (2004) *Risk Management and Insurance*. Second edn. Chicago, IL: McGraw-Hill.

HELCOM. (2013) *Chemical Munitions Dumped in the Baltic Sea: Report of the ad hoc Expert Group to Update and Review the Existing Information on Dumped Chemical Munitions in the Baltic Sea* (HELCOM MUNI). Baltic Sea Environment Proceeding (BSEP) No. 142.

Helmreich, R. L. and Foushee, H.C. (2010) Why CRM? Empirical and Theoretical Bases of Human Factors Training. In Kanki, B.G., Helmreich, R.L. and Anca, J. (eds) *Crew Resource Management*. Second edn. Amsterdam: Elsevier, pp. 3–57.

Hufnagl, P. (2013) Review of Monitoring Techniques for Biological Contaminants. European Commission, JRC, Report EUR 26495 EN, December.

Hughes, S. (n.d.) Cost-Effective Application of the ALARP Principle. Available at: http://barrington.cranfield.ac.uk/resources/20%20Hughes%20-%20Cost%20Effective%20Application%20Of%20The%20ALARP%20Priciple.pdf

Hurley, D. (2000) *Critical Infrastructure Protection: Roles of the Government and Private Sectors and Opportunities for Those Who Carry Out These Roles.* U.S. Department of Commerce, National Telecommunications and Information Administration.

INSAG. (1988) Summary Report on the Post-Accident Review Meeting on the Chernobyl Accident. A Report by the International Nuclear Safety Group. INSAG Series No. 1. Vienna: IAEA.

INSAG. (1991) Safety Culture, Safety Series No. 75-INSAG-4. A Report by the International Nuclear Safety Advisory Group, Vienna: IAEA.

INSAG. (1992) The Chernobyl Accident: Updating of INSAG-1, Safety Series No. 75-INSAG-7A. Report by the International Nuclear Safety Advisory Group, Vienna: IAEA.

IRGC. (2005) Critical Infrastructures: Fact Sheet. International Risk Governance Council.

ISO. (2009a) Risk Management – Principles and guidelines. ISO 31000.

ISO. (2009b) Risk Management – Vocabulary – Guidelines for Use in Standards. ISO Guide 73:2009.

ISO/IEC. (2009) Risk Management – Risk Assessment Techniques. IEC/FDIS 31010.

ISO/PAS. (2007) Societal Security – Guideline for Incident Preparedness and Operational Continuity Management. International Organization for Standardization 22399:2007 (now withdrawn).

Jachia, L. and Nikonov, V. (n.d.) Application of Risk-based Management System Standards to the Design of Regulatory Systems. Available at: http://www.unece.org/fileadmin/DAM/trade/wp6/documents/ManagementSystemStandardsForRegulatorySystems.pdf

Jones-Lee, M. and Aven, T. (2011) ALARP – What does it really mean? *Reliability Engineering and System Safety*, 96, pp. 877–882.

Khan, F. I., Haddara, M. M. and Bhattacharya, S. K. (2006) Risk-Based Integrity and Inspection Modeling (RBIIM) of Process Components/System. *Risk Analysis*, 26(1), pp. 203–221.

Kletz, T. A. (2005) Looking Beyond ALARP. Overcoming its Limitations. *Process Safety and Environmental Protection*, 83(B2), pp. 81–84.

Klinke, A. and Renn, O. (2002) Precautionary Principle and Discursive Strategies: Classifying and Managing Risks. *Journal of Risk Research,* 4(2), pp. 159–174.

Kriebel, D. et al. (2001) The Precautionary Principle in Environmental Science. *Environmental Health Perspectives*, 109(9), pp. 871–876.

Lagadec, P. (1997) Learning Processes for Crisis Management in Complex Organizations. *Journal of Contingencies and Crisis Management*, 5(1), pp. 24–31.

La Porte, T.R. (1994) A Strawman Speaks Up: Comments on the Limits of Safety. *Journal of Contingencies and Crisis Management*, 2(4), pp. 207–211.

La Porte, T.R. and Consolini, P. (1998) Theoretical and Operational Challenges of "High-Reliability Organizations": Air-Traffic Control and Aircraft Carriers. *International Journal of Public Administration*, 21(6–8), pp. 847–852.

Lee, T.R. (1996) Perceptions, Attitudes and Behaviour: The Vital Elements of a Safety Culture. *Health and Safety*, pp. 1–15.

Lemen, R.A. (2004) Asbestos in Brakes: Exposure and Risk of Disease. *American Journal of Industrial Medicine*, 45, pp. 229–237.

Llorca, M. and Rodríguez-Mozaz, S. (2013) State-of-the-Art of Screening Methods for the Rapid Identification of Chemicals in Drinking Water. European Commission, JRC, Report EUR 26643 EN, June.

Majone, G. (2010). Strategic Issues in Risk Regulation and Risk Management. In: OECD Reviews of Regulatory Reform. OECD Publishing. Available at: www.oecd.org/publications/risk-and-regulatory-policy-9789264082939-en.htm

McEntire, D.A. (2015) *Disaster Response and Recovery: Strategies and Tactics for Resilience*. Hoboken, NJ: John Wiley & Sons.

Melchers, R.E. (2001) On the ALARP Approach to Risk Management. *Reliability Engineering and System Safety*, 71, pp. 201–208.

Morris, J. (ed.). (2000) *Rethinking Risk and the Precautionary Principle*. Oxford: Butterworth-Heinemann.

Moubray, J. (1997) *Reliability-centred Maintenance*. Oxford: Butterworth/Heinemann.

Moulton, B. and Forrest, Y. (2005) Accidents Will Happen: Safety-Critical Knowledge and Automated Control Systems. *New Technology, Work and Employment*, 20(2), pp. 102–114.

Moura, M.D.C., Lins, I.D., Droguett, E.L., Soares, R.F. and Pascual, R. (2015) A Multi-Objective Genetic Algorithm for Determining Efficient Risk-Based Inspection Programs. *Reliability Engineering and System Safety*, 133(0), pp. 253–265.

Muchiri, P., Pintelon, L., Gelders, L. and Martin, H. (2011) Development of Maintenance Function Performance Measurement Framework and Indicators. *International Journal of Production Economics*, 131(1), pp. 295–302.

Mussington, D. (2002) Concepts for Enhancing Critical Infrastructure Protection. Relating Y2K to CIP Research and Development. Prepared for the Office of Science and Technology Policy, RAND Science and Technology Policy Institute.

Newsome, B. (2015) The 6.5 Ts: Rationalizing and Managing Risk Management Strategies. *International Journal of Risk Assessment and Management*, 18(1), pp. 89–104.

Nollkaemper, A. (1996) "What You Risk Reveals What You Value", and Other Dilemmas Encountered in the Legal Assaults on Risks. In Freestone, D. and Hey, E. (eds) *The Precautionary Principle and International Law*. Dordrecht: Kluwer Law International, pp. 73–94.

OECD. (2010) Risk and Regulatory Policy. Improving the Governance of Risk. *OECD Reviews of Regulatory Reform*. OECD Publishing. Available at: http://www.oecd.org/publications/risk-and-regulatory-policy-9789264082939-en.htm

Parker, C.F. and Stern, E.K. (2002) Blindsided? September 11 and the Origins of Strategic Surprise. *Political Psychology*, 23(3), pp. 601–630.

Parker, D., Lawrie, M. and Hudson, P. (2006) A Framework for Understanding the Development of Organisational Safety Culture. *Safety Science*, 44, pp. 551–562.

PDD. (1998) The Clinton Administration's Policy on Critical Infrastructure Protection: Presidential Decision Directive 63 (PDD 63), May 22.

Pearce, D. (1994) The Precautionary Principle in Economic Analysis. In O'Riordan, T. and Cameron, J. (eds) *Interpreting the Precautionary Principle*. London: Earthscan, pp. 132–151.

Perrow, C. (1984) *Normal Accidents: Living with High-Risk Technologies*. New York: Basic Books.

Perrow, C. (1994) The Limits of Safety: The Enhancement of a Theory of Accidents. *Journal of Contingencies and Crisis Management*, 2(4), pp. 212–220.

Pritchard, C.L. (2015) *Risk Management. Concepts and Guidance*. Fifth edn. Boca Raton, FL: CRC Press.

Purdy, G. (2010) ISO 31000:2009 – Setting a New Standard for Risk Management. *Risk Analysis*, 30(6), pp. 881–886.

Pursiainen, C. (2009) The Challenges for European Critical Infrastructure Protection. *Journal of European Integration*, (31)6, pp. 721–739.

Pursiainen, C. and Gattinesi, P. (2014) *Towards Testing Critical Infrastructure Resilience*. JRC Scientific and Policy Reports. Luxembourg: Publications Office of the European Union.

Raheja, D., Linas, J., Nagi, R. and Romanowski, C. (2006) Data Fusion/Data Mining-Based Architecture for Condition-Based Maintenance. *International Journal of Production Research*, 44(14), pp. 2869–2887.

Raich, J. (2013) Review of Sensors to Monitor Water Quality. European Commission, JRC, Report EUR 26325 EN, December.

Reason, J. (1990) *Human Error*. Cambridge: Cambridge University Press.

Reason, J. (1997) *Managing the Risks of Organisational Accidents*. Aldershot: Ashgate.

Robinson, P., Woodard, J. and Varnado, S. (1998) Critical Infrastructure: Interlinked and Vulnerable. *Issues in Science and Technology*, 15, pp. 61–67.

Saeed, S., Bajwa, I. S.. and Bakwa, Z. (2014) *Human Factors in Software Development and Design*. Advances in Systems Analysis, Software Engineering, and High Performance Computing (ASASEHPC) Book Series.

Salas, E., Prince, C., Baker, D.P. and Shrestha, L. (1995) Situation Awareness in Team Performance: Implications for Measurement and Training. *Human Factors*, 37(1), pp. 123–136.

Sandin, P. (1999) Dimensions of the Precautionary Principle. *Human and Ecological Risk Assessment: An International Journal*, 5(5), pp. 889–907.

Sandin, P., Peterson, M., Hansson, S.O. and Juthe, A. (2002) Five Charges Against the Precautionary Principle. *Journal of Risk Research*, 5(4), pp. 287–299.

Sargent, R.P., Shepard, R.M. and Glantz, S.A. (2004) Reduced Incidence of Admissions for Myocardial Infarction Associated with Public Smoking Ban: Before and After Study. *BMJ*, 328(977).

Senders, J.W. and Moray, N.P. (1991) *Human Error: Cause, Prediction, and Reduction*. Hillsdale, NJ: Lawrence Erlbaum Associates, Publishers.

Stanton, N.A. and Edworthy, J. (1999) Key Topics in Auditory Warnings. In Stanton, N.A. and Edworthy, J. (eds) *Human Factors in Auditory Warnings*. Aldershot: Ashgate, pp. 345–359.

Sunstein, C.R. (2003) Beyond the Precautionary Principle. *University of Pennsylvania Law Review*, 151(3), pp. 1003–1058.

Talbot, J. (n.d.) ALARP (As Low As Reasonably Practicable). Available at: http://www.jakeman.com.au/media/knowledge-bank/alarp-as-low-as-reasonably-practicable

Tanchou, V. (2014) Review of Methods for the Rapid Identification of Pathogens in Water Samples. European Commission, JRC, Report EUR 26881 EN, September.

UNISDR. (2009) *UNISDR Terminology on Disaster Risk Reduction*. United Nations International Strategy for Disaster Reduction (UNISDR), Geneva, Switzerland, May 2009. Available at: http://www.unisdr.org/we/inform/terminology

Van Asselt, M.B.A. and Renn, O. (2011) Risk Governance. *Journal of Risk Research*, 14(4), pp. 431–449.

Van Asselt, M.B.A. and Vos, E. (2006) The Precautionary Principle and the Uncertainty Paradox. *Journal of Risk Research*, 9(4), pp. 313–336.

Vecchio-Sudus, A.M. and Griffiths, S. (2004) Marketing Strategies for Enhancing Safety Culture. *Safety Science*, 42, pp. 601–619.

Weick, K.E. (1987) Organizational Culture as a Source of High Reliability. *California Management Review* 29(2), pp. 112–127.

Weick, K.E. (1988) Enacted Sensemaking in Crisis Situations. *Journal of Management Studies*, 25, pp. 305–317.

Weick, K. E., Sutcliffe, K. M. and Obstfeld, D. (1999) Organizing for High Reliability: Processes of Collective Mindfulness. In Sutton, R.S. and Staw, B.M. (eds) *Research in Organizational Behavior, Volume I*. Stanford, CA: JAI Press, pp. 81–123.

Westrum, R. (1996) Human Factors Experts Beginning to Focus on Organizational Factors in Safety. *ICAO Journal*, 8, pp. 6–8, 26–27.

Westrum, R. (2004) A Typology of Organisational Cultures. *Quality and Safety in Health Care*, 13(Suppl 2), pp. 22–27.

Wickens, C.D. (2005) Attentional Tunneling and Task Management. Technical Report, AHFD-05-23/NASA-05-10, December. Prepared for NASA. Ames Research Center, Moffett Field, CA Contract NASA NAG 2-1535, Aviation Human Factors Division, Institute of Aviation, University of Illinois.

Wilkinson, G. and David, R. (2009) Back to Basics: Risk Matrices and ALARP. In Dale, C. and Anderson, T. (eds) *Safety-Critical Systems: Problems, Process and Practice*, London: Springer-Verlag, pp. 179–182.

Wittneben, B.B.F. (2012) The Impact of the Fukushima Nuclear Accident on European Energy Policy. *Environmental Science & Policy*, 15(1), pp. 1–3.

Wu, T.-C., Lin, C.-H., and Shiau, S.-Y. (2010) Predicting Safety Culture: The Roles of Employer, Operations Manager and Safety Professional. *Journal of Safety Research*, 41, pp. 423–431.

Yoe, C. (2012) *Primer on Risk Analysis: Decision Making Under Uncertainty*. Boca Raton, FL: CRC Press.

Zhu, A.T., von Zedtwitz, M., Assimakopoulos, D. and Fernandes, K. (2016) The Impact of Organizational Culture on Concurrent Engineering, Design-For-Safety, and Product Safety Performance. *International Journal of Production Economics*, 176, pp. 69–81.

Zohar, D. (1980) Safety Climate in Industrial Organizations: Theoretical and Applied Implications. *Journal of Applied Psychology*, 65(1), pp. 96–102.

Zohar, D. (2000) A Group-Level Model of Safety Climate: Testing the Effect of Group Climate on Microaccidents in Manufacturing Jobs. *Journal of Applied Psychology*, 85(4), pp. 587–596.

4 Preparedness

If crises cannot be prevented, one should at least be prepared for them. This preparedness should be based on the risk treatment results. The preparedness efforts should be directed towards those risks that were selected as risk treatment-worthy from the organisation's perspective but that could not be completely eliminated with the efforts discussed in the previous chapter. Thus, in contrast to the subject area of Chapter 3, the occurrence of a crisis is assumed. However, preparedness might help to mitigate the consequences of a crisis when it hits. In that sense, good preparedness and its individual components can be seen as the continuous building of consequence-reducing barriers, and hence the sharp demarcation between prevention and mitigation efforts and preparedness disappears.

A good, if rather generic point of departure for defining preparedness is that it consists of the knowledge and capacities developed by all kinds of organisations and individuals to effectively anticipate, respond to and recover from the impacts of likely, imminent or current hazard events or conditions (cf. UNISDR, 2009). There are also many other concepts that come close to that of preparedness, such as readiness. Indeed, it has been noted that "preparedness is the state of readiness" in the context of disaster management (Perry and Lindell, 2003, p. 338). Another related concept is resilience, popularised in the 2000s in particular. Boin and Lagadec even argued in their 2000 article, somewhat ahead of most scholars, that "the future crises require a preparatory effort that contains both resilient-oriented strategies and anticipation-based strategies", and that "resilience is the key to future coping" with crises (2000, p. 188). Some 12 years later, Hémond and Robert (2012) argued that a paradigm shift could indeed be identified from striving for a state of preparedness towards a state of resilience.

It is easy to agree that the concept of resilience is the new catchword in all risk-related debates. It is also clear that it overlaps with preparedness, albeit constituting a much broader concept. A generic definition, applicable to almost all types of systems and organisations, is provided by UNISDR (2009):

> The ability of a system, community or society exposed to hazards to resist, absorb, accommodate to and recover from the effects of a hazard in a timely and efficient manner, including through the preservation and restoration of its essential basic structures and functions.

We will discuss the concept of resilience in some detail in Chapter 6 when focusing on recovery, which, in a way, is at the very core of the resilience concept. However, in this chapter's context, one should note that "resist, absorb, accommodate to and recover" actually mean that resilience can be understood as an umbrella concept including measures covering basically all of the different phases and elements of the crisis management cycle. Hence, the problem with resilience as a replacement for preparedness is that in order to understand what it means and how it could be operationalised, one has to divide it into sub-themes or indicators. So, in this process one returns to the concept of preparedness once again.

What then are the sub-themes, components or elements that belong to the concept of preparedness? The UNISDR (2009) definition states that preparedness is based on a sound analysis of disaster risks and good linkages with early warning systems, and includes such activities as "contingency planning, stockpiling of equipment and supplies, the development of arrangements for coordination, evacuation and public information, and associated training and exercises", which must be supported by formal institutional, legal and budgetary capacities. Haddow, Bullock and Coppola (2011, p. 102) propose "the four major components of any preparedness effort: planning, equipment, training, and exercise". Drennan, McConnell and Stark (2015, p. 134) suggest four "main stages" of pre-crisis contingency planning: threat assessment and scenario creation; contingency planning and resource dedication; testing through simulations and exercises; and capability assessment and audits. The stage at which a crisis manager might begin the cycle depends on the history of the respective organisation in terms of its risk and crisis management practice. Lunde (2014, pp. 51, 52) in turn looks at preparedness from three perspectives. Operational preparedness is about what should be done and how. Organisational preparedness defines who should be doing what and when, while technical preparedness contributes with supporting tools and equipment. Radvanovsky and McDougall (2010), for their part, subdivide preparedness into planning, training, exercises, personnel qualification and certification, equipment acquisition and certification, mutual aid agreements and publication management.

To concretise this, say, in the case of electricity supply disruptions, might point to a wide range of preparatory activities: operators having ready-made plans for how and when personnel should be called in, or put on stand-by; keeping maps up-to-date; maintaining information about how disruption could affect operations; monitoring weather forecasts; making arrangements with third parties for providing spare parts and additional equipment; preparing for cooperation with the emergency services; pinpointing vulnerabilities in telecommunication nodes, waterworks and sewage farms; or prioritising support for vulnerable groups such as hospitals, nursing homes for the elderly, schools, daycare centres, and so on (Landstedt and Holmström, 2007).

While any of the above taxonomies would provide a good starting point for discussing the concept of crisis preparedness, we combine these facets to capture the main components of this multidimensional concept. In so doing, we discuss in some detail the following preparedness domains: planning;

organisation and procedures; capacity-building; redundancy; agreements and pre-arrangements; capability-building; and early warning.

4.1 Planning

Planning is the core of preparedness, be it a question of organisations dealing with financial crises (e.g. Harvard, 2004, pp. 35–51), disaster management (e.g. McConnell and Drennan, 2006), hospital readiness (e.g. Power, 2008) or any other types of organisations, and even at the household and individual level (e.g. Paton, 2003). As Fink (2002, p. 54) puts it:

> Every business, large or small, public or private, should have a crisis management plan. Every division of every company, industrial or service business, should also have a crisis management plan. There are no exceptions, merely differences in degrees.… There is no room for complacency and no excuse for a lack of planning.

While there might be cultural differences – some countries are better at crisis management than others, and high-risk industries are naturally more risk-aware – large-scale surveys indicate that the larger the organisation, the more likely it is to have a crisis plan (Johansen, Aggerholm and Frandsen, 2012).

Yet preparedness issues tend to be given low priority and receive little attention until a crisis actually occurs, as noted by Enander, Hede and Lajksjö (2015) when focusing on municipality-level crisis management. The same applies to business. Carmeli and Schauerbroeck (2008) have noted, based on surveys, that crisis events are often critical turning points in a business's life and reveal unknown problems in organisational design and practices. The need for attention to these problematic areas often only emerges in times of crisis, when it is too late to start developing a viable crisis management system. Many organisations are ill-prepared for critical situations, and often respond to such situations in ways that made these crisis events worse. Organisations are then often faced with substantial costs that could have been avoided or reduced, had these businesses been better prepared to manage the ensuing crisis.

Crises are low-probability events that are perceived as unlikely, and the benefits of crisis preparedness are less obvious in comparison with other areas of public expenditure, such as health or education, or in the business world on efficiency, for instance. Moreover, preparedness resources tend to be directed more towards focusing on the last crisis rather than on uncertain future problems. McConnell and Drennan (2006) have argued as much about public institutions in general. In addition to seeing crises as low-probability events, they also blame the fact that the resources and substantial time needed for crisis preparedness are not regarded as providing worthwhile payoffs. Furthermore, Boin and 't Hart (2003) have noted that while the general public expects leaders to put safety first, leaders consider the economic and political costs of regulating and enforcing maximum safety to be too expensive.

Governments and leaders are often reluctant to prepare themselves for their crisis-response role, and they also misinterpret or ignore repeated indications of impending danger. When a crisis hits, the management and decision-making are usually too top-down control and command-oriented, while much more multi-organisational coordination would generally be demanded. Added to this, leaders often fail to learn from the crises. While Piotrowski (2006) has further identified differences in the crisis management and preparedness levels between individual businesses, local public institutions (such as the police) and national public organisations, Parnell (2015) has noted variations between different countries. On the other hand, high-risk or high-reliability industries and organisations often take preparedness measures that exceed their international and national legal obligations (e.g. Tammepuu, Tammepuu and Sepp, 2009; McConnell and Drennan, 2006, pp. 67–69).

Good crisis preparedness starts with and presupposes what is usually called contingency planning, namely, a management process that analyses specific potential events or emerging situations and establishes arrangements such as capacity- and capability-building in advance, in order to enable timely, effective and appropriate responses to such events and situations.

The need for preparedness planning is rooted in the very definition of a crisis, as presented in Chapter 1. Time is an important currency in a crisis situation, and the problem is precisely that one usually has too little of it. There is a limited time for decision-making, due to an approaching deadline or the growing costs of inactivity. Crisis decision-making is therefore necessarily limited in terms of rationality, due to uncertainty and limited time to consider all the options. The point of contingency planning is to make as many decisions as possible before a crisis occurs. Pre-crisis planning provides some extra room to consider a wide range of options in advance and even to test them against the chosen crisis scenarios.

On the other hand, Pinkowski (2009) has warned about the "delusion of preparedness" and noted that preparedness planning may ultimately be fortuitous and a hunch at best. McConnell and Drennan (2006, p. 64) have pointed out that "it is difficult to produce a single plan that covers all the potential challenges emerging from crisis situations". Stern (2013) has emphasised that contingency planning should actually be about planning how to improvise. In any case, planning should not be too rigid and should not be an obstacle to improvisation.

A look at the literature on preparedness soon reveals that while the multi-disciplinary literature is huge, especially in the field of disaster preparedness and all the technologies related to it, the field is nonetheless rather under-theorised. This is also true of preparedness planning. However, in recent decades one has been able to identify a general discussion on planning theory in other fields, perhaps most active within the field of urban planning. One of the main theoretical debates has been that between instrumental and communicative planning (e.g. Sager, 2013), reminiscent of the risk management vs. risk governance debate briefly discussed at the end of Chapter 2. Instrumental planning is based on the idea of rationality, the goal residing in the search for a connection or optimal balance between the goal and the means. In communicative planning, the focus is on dialogue between a variety of stakeholders in order to agree on a plan, or at least to let all stakeholders have a say.

When we apply this notion to crisis preparedness planning, the planning paradigm one uses depends largely on the context. If an individual organisation, such as a business corporation, is planning its preparedness against different types of hazards, it may benefit from instrumental planning. If a community, such as a municipality with many immediate stakeholders, is planning measures to prepare itself for emergencies and disasters, it may well include as many stakeholders in the process as possible. However, this distinction should not be too rigid. Organisations of all types should consider not only internal resources and processes, but also coordination and communication with external stakeholders or society at large.

Most of the literature on contingency planning comprises case studies, however, and is only occasionally guided by any clearly articulated theoretical framework. Yet some normative principles and taxonomies can be found that make it easier to deal with this rather multidimensional concept and define its practical applications. One of the early efforts was Dynes, Quarantelli and Krep's (1972) article on disaster preparedness from a community perspective, followed by Quarantelli's (1984) somewhat more developed "general principles of disaster preparedness planning". Even the latter was actually a short practical checklist for community disaster response organisations. This kind of approach is helpful in identifying and evaluating preparedness plans, especially those of communities and municipalities (see Burlin and Hyle, 1997). One can also find rather early but nonetheless sophisticated checklists aimed at any organisation (e.g. Pearson and Mitroff, 1993).

By and large, the principles and recommendations remained rather abstract, however. Alexander (2005; cf. Perry and Lindell, 2003) summarised the literature on the criteria for an emergency planning standard into "18 axioms". These include generic statements such as recommendations that the plan should be prepared by, or under the direction of, a qualified emergency planner; should be written in clear, simple, unambiguous language; should conform to the laws on emergency and disaster management; or that the plan should be based on a careful and, as far as possible, exhaustive assessment of what is likely to happen when an emergency occurs. Somewhat more detailed and measurable preparedness evaluation methodologies have subsequently been developed (e.g. Simpson, 2008), with some indicators of good planning identified and then applied to case studies where these indicators have been scored. Pearson and Mitroff (1993) divided their checklist into five action sets: strategic; technical and structural; evaluation and diagnostic; communication; and psychological and cultural. Each action set has half a dozen normative subprinciples, such as recommendations to dedicate budget expenditures for crisis management, ensure technological redundancy in vital areas, establish a tracking system for early warning signals, improve communication lines with external stakeholders, and provide psychological support services.

74 *Preparedness*

Basically, anticipation of problems, issues and challenges is a central element of contingency planning. This naturally depends on the context. In the field of disaster management, for instance, this might include problems in communication, evacuation, transportation, coordination, and so forth (Veenema and Woolsey, 2012, p. 8; Quarantelli, 1984). Kaneberg, Hertz and Jensen (2016) in turn discuss emergency management planning in terms of a "supply chain". They divide the objects of planning into five tasks. First, human resources, involving selecting and training the people who are supposed to have a role in emergency response. Second, knowledge management, referring to the knowledge gained from previous emergencies, which has to be utilised in preparedness planning. Third, planning, including ensuring resources are available and identifying alternative resources for a quick and effective response. Fourth, presupposing that financial resources or arrangements have been ensured in advance. Finally, planning how to coordinate the cooperation and communication with civil society. Drennan, McConnell and Stark (2015, p. 134) speak about a planning cycle, consisting of assessing threats and creating scenarios, developing a contingency plan and dedicating resources, testing the plan through simulations and exercises, and assessing the capabilities added through an audit.

Efforts have been made to standardise contingency planning. The International Organization for Standardization (ISO) has published ISO 22301, namely, guidelines and standards on preparedness and business continuity management (ISO, 2012; see also ISO/PAC, 2007) to this end. In the same vein as the risk management standard, this standard also starts by determining the internal and external context. It goes on to discuss the importance of top management commitment and responsibility, the importance of integrating contingency planning into the organisation's strategic goals and day-to-day practices, gives some guidance on implementation, and underlines that the plan must be evaluated and continuously updated. There is also an ISO 41000 family of standards (ISO, 2015) for environmental management systems, where emergency preparedness plays a central role. Like most ISO standards, these are also applicable to any organisation, regardless of size, type and nature, and apply to the respective aspects of its activities, products and services that the organisation determines it can either control or influence. Organisations can seek certification to verify that they are following the standard.

The ISO 2800 family of standards, focusing on supply chain security management (ISO, 2007, 2011, 2014a–c), is suitable for many industries that deal with critical infrastructures, such as ports. Separate guidelines are prepared for small and medium businesses. The standards themselves can be seen as checklists for preparing contingency plans, especially against security threats, although the main standard has a specific section on what should be included in the "emergency response and recovery plan".

A standard rather similar to ISO 22301 but with more detailed guidelines is provided by the US National Institute of Technology and Standardization. While it focuses only on federal (US) information systems contingency planning (NIST, 2010), it includes rather generic principles. The guidelines define a seven-step contingency planning process, which also covers risk analysis and prevention. First, the organisation in question should develop a contingency planning policy statement to give legitimacy to the preparedness efforts. Second, it should conduct a business impact analysis in order to identify and prioritise systems and components critical to supporting the organisation's mission or business processes. Third, the organisation should identify preventive controls or barriers. Fourth, it should create contingency and recovery strategies to ensure that the system may be recovered quickly and effectively following a disruption. Fifth, the organisation should develop the contingency plan itself, namely, detailed guidelines and procedures for restoring a damaged system. Sixth, this should be followed by testing the plan, usually with exercises. Finally, the plan's maintenance should be ensured; it should be a living document that is updated regularly.

Similar types of guidelines for contingency planning can be found in connection with many other public bodies, companies, international organisations and non-governmental organisations. The International Federation of Red Cross contingency planning guidelines (IFRC, 2012), for instance, break contingency planning down into five main steps: prepare, analyse, develop, implement and review. In more simple terms, they ask, "what is going to happen, what are we going to do about it, what can we do ahead of time to get prepared?", emphasising that contingency planning is an ongoing process and the planning process is often as important as the plan itself.

The general philosophy of contingency planning tends to emphasise integrated approaches and coordination, including not only different public institutions at national and local levels, but also the private sector and civil society. McConnell and Drennan (2006), among others, have nonetheless pointed out that while this ideal form of integrated approach can be defended, crisis management is often different in reality. Hence, a certain tension remains over the ownership of contingency planning.

If one looks at existing contingency plans, one quickly sees that while they are usually based on the very issues and principles outlined above, they differ considerably in scope. This is quite natural since such plans necessarily depend on their application area and the respective organisation, and their value often lies in how much detail they go into in their presentation of pre-planned procedures and the respective resources. In many countries, communities or municipalities have a legal obligation to produce a contingency plan. In Norway, for instance, 94 per cent of municipalities have fulfilled this obligation to date. According to the national guidelines, the plans should at least include a crisis leadership plan, a plan for communication, a review of resources, alert and warning lists with contact information, and evacuation plans. Many of the plans also include local risk scenarios (DSB, 2016). Such plans are usually rather brief, providing local information about evacuation locations. Authorities working with safety and security provide more comprehensive plans. The contingency plan of the Norwegian police, for instance, is an impressive document (PBS, 2011), with the first part alone comprising about 300 pages. It covers most areas that are typically found in a contingency plan. Issues such as basic concepts, leadership and responsibilities at strategic, tactical and operational levels, own and external resources, coordination with external domestic and international actors, internal and external communication (including media management), warning systems, and the specific crisis areas or scenarios one is preparing for are all considered in detail. In many cases, national authorities have also prepared contingency plans focusing on only a limited crisis or disaster, such as the Swedish contingency plan for nuclear accidents (MSB, 2016).

4.2 Organisation and procedures

One of the most important functions of the contingency plan is to define the responsibilities, roles and tasks of the staff, parts of the organisation and, in the case of a larger set of organisations, the respective responsibilities of the relevant organisations in a crisis situation in a multidisciplinary, cross-sectorial and multi-jurisdictional environment. But how does one know who to appoint as the responsible persons in the case of unexpected events and crises whose characteristics remain unknown? Should one have a hierarchical or more flexible organisation in a crisis?

When it comes to crisis management, China, for example, relies on a rather unified so-called control and command system. True, the emergency response organisation is defined according to the severity and geographic scope of the crisis, extending from county and municipal levels to provincial and state levels. In any case, unified command is applied. If the responsibility remains at the local level, for instance, "units and departments within certain areas differ in terms of administrative subordinate relations, but under a state of emergency all obey the unified command of the local government" (Fujin and Tiehan, 2008, pp. 7, 48). In many smaller countries' emergency management systems, such as the Netherlands, rather strict unified control and command systems are also applied, combining strategic and operational crisis leadership hierarchically (Scholtens, Jorritsma and Helsloot, 2014). The Scandinavian version of the control and command system, offering modest flexibility and ad hoc members in crisis management, follows three principles in its contingency planning, namely, the principles of responsibility, equality and proximity (e.g. Lunde, 2014). The principle of responsibility means that the person in charge of a field in a normal situation also has the responsibility to deal with unwanted or extraordinary events and crises in the same area. The principle of equality means that the crisis organisation should be as similar as possible to the organisation operating in normal conditions. Subsidiarity means that adverse or extraordinary events and crises should be handled organisationally at the lowest possible level.

Additionally, the classical view is that crisis management requires a dedicated crisis management group, led by the highest possible official of the respective organisation, which in a way enforces the control and command system. Standards (e.g. ISO 2007) also emphasise the importance of the participation of top management. In standardised guidelines too, the normative rule is typically that roles and responsibilities should be defined in advance. This rule is also compatible with the control and command system, which was originally borrowed from the military. The command system can be divided into different, smaller groups, such as operations, planning and intelligence, logistics, communication, and finances and administration (Heath, 1994, pp. 142, 143). Watters (2014, pp. 247–251, app. F), in turn, proposes a blueprint for any organisation's crisis management team, consisting of a crisis leader, business continuity manager, administrative support, emergency service liaison, facility managers, communications managers, human resources and information technology managers, each having pre-defined responsibilities.

More recently, however, the control and command leadership model has been heavily criticised as being too rigid and hierarchical (e.g. Scholtens, Jorritsma and Helsloot, 2014; Groenendaal, Helsloot and Scholtens, 2013). One should rely instead on a more flexible approach, allowing ad hoc structures or team members. This might be considered important due to the specific nature of the crisis. The weakness of these types of ad hoc crisis groups, however, is that this practice may demotivate those who deal with the same issue area on a daily basis, thus potentially undermining the efforts of pre-crisis preventive and preparative actions. Ad hoc groups are also difficult to train to work together, and they may bring new conflicts into the work (e.g. Heath, 1994, pp. 142, 143).

A model that is ostensibly somewhere in between has been called the network management model, presupposing a rather open, horizontally organised decision-making approach. This kind of decision-making demands some kind of coordination and negotiation centre or body, pre-arrangements concerning obligations and responsibilities, and trust (Drennan, McConnell and Stark, 2015, pp. 174–175). The US has explicitly experienced the need to develop new organisations and frameworks to ensure that its crisis response is both well coordinated and flexible. Such institutions as the National Response Framework and the National Incident Management System are both aimed at coordinating the activities of federal, state and local levels of administration in a multiagency spirit with the private sector and the voluntary sector (Radvanovsky and McDougall, 2010).

There is also the question of the level at which the responsibility in a crisis should lie. A study of 22 European countries' civil protection systems revealed that they follow largely different leadership systems. A few countries follow a decentralised model, with most being either rather decentralised, rather centralised or centralised (Kuipers et al., 2015; cf. Pursiainen, Hedin and Hellenberg, 2005). In the crisis management literature, centralised and decentralised response structures usually refer to different decision-making levels such as the national government, regional authorities or municipality (rather than the size of the group, which is sometimes the case when discussing political crisis management). The question then concerns whether the decisions are made at the central government level or by local authorities, for instance. The Scandinavian solution to this question was presented above. There is often a normative rule involved in discussing this issue. A centralised response would mean that the majority of operational and strategic decisions are taken by a central group of high-level decision-makers, and this would be effective when a crisis response calls for additional authority and resources. While most crisis situations require some form of front-line response (e.g. police, rescuers, healthcare workers), a truly decentralised response in turn would be based around the principle of subsidiarity, taking decisions at the lowest possible level. This would be effective when a crisis response needs to be tailored to local circumstances, for which the necessary vital information lies only at that level of administration, obviating the need for large-scale coordination that would extend far beyond the local community.

While there is no consensus on the best model of organisation, and the practical applications are often hybrid combinations of both command and flexible systems, this same ambiguity can be extended to procedures that are planned for crisis response. Organisations often rely on standard operating procedures. These may be divided into several subgroups, but basically call for understanding the roles, responsibilities and jurisdictions of different actors in a certain crisis situation, as well as the equipment and other resources available from other organisations. Procedures often include rather detailed guidelines on such concrete issues as who is responsible for what, how to coordinate cooperation with other actors, how to organise communication management or site management of a disaster scene, how to establish a command post or deal with media relations, how to minimise the consequences of an unexpected event, how to document the lessons learned, and so forth. This usually involves using ready-made protocols and templates (cf. e.g. Radvanovsky and McDougall, 2010, pp. 109–134).

Standardised procedures are generally regarded as being characteristic of good administration, which is why their development is often on the agenda for crisis management purposes. In more theoretical terms, the so-called bounded rationality school (e.g. Lindblom, 1959; Simon, 1972a, 1972b; Bendor, 2010a) emphasised some decades ago that organisational routines would compensate for the problems caused by the lack of rationalism due to the cognitive limits of an individual decision-maker. However, Allison (1971), in his classic book on high politics crisis decision-making, has presented the so-called organisation model based on precisely the opposite idea: that organisational routines may limit rationality (Bendor, 2010b). Allison would probably agree that in normal circumstances standard operational procedures are useful in adding efficiency and rationality, but his argument was that organisations do not always see efficiency as the ultimate goal; rather their activity is guided by the organisation's identity and administrative culture, or simply the organisation's prestige (Allison and Zelikow, 1999, pp. 147–153). In a crisis, organisational routines may become an obstacle to considering and finding flexible and creative solutions that the situation might lend itself to. Thus, in Allison's scheme, organisational routines may cause non-rational behaviour. In contrast, however, it has also been noted that the breakdown of organisational routines, especially in a crisis situation, may become a bottleneck in the administrative and decision-making system, and that any such system would be virtually impossible without standardised procedures (Bendor and Hammond, 1992, p. 313). In other words, crisis management and decision-making are also based on rules, norms and institutionalised practices and plans as well as existing resources, which both enable as well as condition and limit the decision-makers.

4.3 Capacity-building

A range of institutional and legal solutions, strategies and policies, processes and tools are naturally needed to define and support attempts to enhance crisis preparedness. However, highly material resources and capacities are also required. "Crises push response organizations into overdrive and require them to mobilize a substantial increase in resources" (Ansell, Boin and Keller, 2010, p. 198). Therefore, it is generally argued that a commitment to crisis preparedness necessitates the respective resources, including monetary resources, equipment and time. Partly for this reason, the commitment must derive from the top management. "They are responsible for deploying resources to promote staff commitment and training. In addition to this, top managers' agreement is necessary for establishing the required technical measures to prevent a crisis from occurring and to absorb the impact" (Labaka, Hernantes and Sarriegi, 2015, p. 99). While financial crises, such as the euro crisis of recent years, would typically call for pre-arranged financial stabilisation mechanisms (Schwarzer, 2015) in order to facilitate crisis management, this is true for almost any kind of crisis. Monetary resources or insurance coverage must be set aside to cover repairs and replacements and possible liability issues that occur just before or after the triggering event. In other words, one should be prepared for crisis response in monetary terms (Labaka, Hernantes and Sarriegi, 2015, p. 100).

In addition to monetary resources, a crisis situation may require certain equipment to be in place when the crisis hits. This may be a question of very simple equipment, which Fink (2002, p. 55), for one, has illustrated with a metaphor describing the need for a torch when a power outage occurs in one's house: "It is during these times when people are apt to get hurt running around in their house because they cannot locate their flashlights." Or it may be a question of very large systems in terms of monitoring, early warning or alert. In any event, equipment clearly needs to be in place in advance in order to manage the crisis response properly. "Equipment is very important in the preparedness phase because it is during this phase that equipment needs are identified, equipment is purchased, and staff is trained in the use of the equipment" (Haddow, Bullock and Coppola, 2011, p. 115).

80   Preparedness

An emergency organisation, national or local, assessing capacities in the case of a mass casualty event, for instance, has to conduct a capacity analysis of hospital and healthcare resources. Hospitals or hospital systems should have planned and invested in their surge capacity in the event of a disaster, including such specific capacities as isolation areas related to biological or radiological terrorism and decontamination capacities (Power, 2008). One consideration is how to maximise the existing local hospital capacity; others include how to use other, more distant hospitals, how to create alternative non-temporary care sites with the necessary infrastructure, and how to ensure that additional professionals and medical volunteers could be utilised alongside regular staff (e.g. Greighton, 2009, pp. 15–16). A nuclear plant has to be prepared to have emergency equipment at hand, both on site and outside the plant, so that the plant is able to absorb the impact and ensure the safety of the workers when a crisis occurs. Naturally, any nuclear plant has a long list of emergency material and equipment, such as communication systems, breathing equipment, special clothes for radiation protection, radiation measurement devices, medical equipment and means of transport (Labaka, Hernantes and Sarriegi, 2015, p. 98). These examples show that there obviously cannot be any blueprint for the list of resources and equipment needed as they will depend on the sector or organisation type as well as the characteristics of the crisis.

A term used in humanitarian and medical crises in particular, especially in the transboundary or international cooperation context, is surge capacity, which covers most of the issues discussed in this section, although it touches on all aspects of preparedness. In the transboundary context, surge capacity usually refers to major emergencies that require (international) organisations to rapidly and effectively increase their resources in terms of people, money and materials in the countries affected by an emergency. It is an ability to scale operations up and down swiftly, smoothly and productively, ensuring that scarce resources are used efficiently and with maximum impact. It presupposes a flexible pre-positioning of funds and equipment before the crisis emerges (Emmens, 2008).

In the chapter on risk assessment, we raised the issue that not all risks can be identified and analysed. They are, therefore, difficult to prevent, mitigate or prepare for. In particular, when it comes to 'imagining the unimaginable' or identifying very unlikely hazards, a way to avoid too much speculation is to start from the consequences. For instance, in the critical infrastructure literature (Pursiainen and Gattinesi, 2014), it has been noted that, due to unexpected dependencies and interdependencies, it is not possible to list all of the functions that are critical in every situation, nor to identify all the possible risks that may lead to a critical infrastructure failure. Therefore, the focus should be on resilience instead, namely, a system's capability to withstand, to adapt to, and to quickly recover from stresses and shocks (European Commission, 2012, p. 5). Hence, while some disasters take place in such complex or chaotic circumstances that cause and effect remain unknown, or appear logical only after the fact, one can always look at risks in terms of their possible consequences. There are indeed many consequences of disasters (such as the need for evacuation) that are to some extent unrelated to the exact root source or nature of the disaster. Therefore, it may be rational to rely on the consequence-oriented definition of criticality (Egan, 2007, p. 5). It is also valuable to use impact assessment to identify the most common consequences, their combinations and interdependencies, thereby working backwards. A consequence-based analysis, concentrating on the impacts instead of the root sources, is particularly useful when the root sources of risks are complex. From the perspective of capacity-building, one could identify the categories of impact that one has to be prepared for, irrespective of the root cause of the crisis.

4.4 Redundancy as preparedness

The need for redundancy has already been mentioned above on many occasions, directly or indirectly. It is a central concept in the broader resilience debate, being a part of preparedness measures, and in some cases of prevention and pre-crisis mitigation measures, as it has to be implemented before the crisis hits. While building redundancy during the crisis response or recovery phase is sometimes possible, relying purely on post-factum solutions is doomed to be less effective than well-prepared and pre-planned redundancy solutions.

Redundancy is a concept mostly discussed in the context of critical systems. From a societal security point of view, the latter include critical infrastructure, usually defined (e.g. European Council, 2008) as an asset, system or part thereof that is essential for the maintenance of vital societal functions such as health, safety, security and the economic or social well-being of people, the disruption or destruction of which would have a significant impact as a result of the failure to maintain those functions. While critical infrastructure is a societal phenomenon in this sense, it is closely related to the organisational level of analysis as many organisations may themselves be a part of the critical infrastructure or at least dependent on the functioning of this critical infrastructure, such as electricity, water distribution or public administration.

In more generic terms, one often speaks about the redundancy of a system, which is one component of the resilient design of systems. Many systems have redundant systems that can replace the original system if a disturbance occurs. This can relate to such issues as organisational arrangements about who is available to take over if one part of an organisation has to be replaced, or where an organisation's activity would be located if the current site became unavailable for some reason due to the crisis event. For the most part, however, the redundancy discourse focuses on rather technological solutions, and is an established discipline within the field of reliability engineering. This fact has somewhat hindered the redundancy debate in other fields as in one sense it is too sophisticated to be applied in more generic crisis management practices, but in another sense is a good starting point as redundancy to some extent becomes a measurable feature of a system.

Redundancy is not only about whether it exists or not. In more precise terms, it emphasises the degree to which the function of a system that is temporarily disrupted can be replaced by other systems, substituted by other solutions, or re-routed, and so forth. This may well be the most important rule of resilient design, namely that an important function should not be dependent on a single element, or that one should not 'put all one's eggs in one basket'. A rather early treatment of the issue of resilient system design by Fiksel (2003, p. 5332) notes that complex, hierarchically organised systems (e.g. aircraft or nuclear plants) tend to have "rigid operating parameters, are resistant to stress only within narrow boundaries, and may be vulnerable to small, unforeseen perturbations". Redundancy would entail that "distributed systems composed of independent yet interactive elements may deliver equivalent or better functionality with greater resilience".

However, the existing literature rarely delves into explaining how this should be carried out in practice, and by nature this is an issue that depends on the particular system or function. In terms of the technological dimension of redundancy, the UK government definition reveals some of the concept's main characteristics and possible solutions for enhancing it. According to the definition, redundancy is the availability of "backup installations or spare capacity" that "will enable operations to be switched or diverted to alternative parts of the network in the event of disruptions to ensure continuity of services" (UK Cabinet Office, n.d., item 2.14). This statement explicitly includes the realistic idea that redundancy strategies would lead to an initial loss of performance until the alternative infrastructure could be brought into operation, but in some sectors the switchover to maintain services might be instantaneous.

The definition also recognises, however, that in some sectors or organisations "it may not always be feasible to operate with significant spare capacity within the network". This negative factor of redundancy, that is, the costs compared to expected benefits, is an often-noted setback. So redundancy is usually only a characteristic of high-risk organisations. NASA (2016), the US National Aeronautics and Space Administration, notes that the application of redundancy "is not without penalties". However, "the costs may be recovered by the increased reliability".

Redundancy is consequently a factor that should be built into a critical system. Basically, it entails the duplication or triplication of the critical elements of a system by means of a backup, with the result that an individual component or function failure, due to a malicious or non-malicious action, would be insufficient to render the system inoperative. When we start to build redundancy, it is sometimes useful to think in terms of counterfactuals. Counterfactual theories of causation (e.g. Menzies, 2014) take as their point of departure the basic idea that the meaning of causal claims can be explained in the form of "If A had not been there, C would not have occurred". While there are several versions within this debate, the theory proposes that non-actual possible worlds should and could be compared with the materialised world in terms of similarity or difference. The idea of a cause is linked to the idea of something that makes a difference from what would have happened if it had been absent. In real life, it is usually a question of a causal chain, although sometimes it may concern a chancy causation without determination.

In the context of redundancy, this means that one should analyse the degree of redundancy the other way around, namely, by considering things that would happen without a backup component in the system. Similarly, while critical systems should have some kind of backup capability for their function or service, "the backup systems or components of the backup systems are most often collocated with the primary systems", or in many cases, "redundant systems feed into a single critical node shared by the primary systems". Therefore, a single electrical distribution panel should not control the flow of commercial power, diesel backup generators and uninterruptible power supply batteries, for instance, and a single manhole should not house all communications lines leading to and from the facility (Baker, n.d.).

How then do we gauge whether the system is redundant or not? This knowledge is usually obtained by planning and testing. While in certain technological fields, like mainstream software development, testing redundancy is usually included, and hence methodologies and methods have been developed in many systems where redundancy is not a planned or tested characteristic. A basic rule would be to create test situations whereby one can evaluate the excess capacity to reduce the impact of component or subsystem failures. This is usually called passive redundancy, often referred to as standby redundancy. Passive redundancy does not always have to be 100 per cent. A good example is human sight. If one loses the sight in one eye, one can continue to rely on the remaining eye, even if the capability of sight is reduced by half. Active redundancy would then focus, for instance, on a situation whereby an overload in one power line and circuit breakers should automatically disconnect this line and redistribute the power across the remaining lines. This raises the question of when to use passive and when to use active redundancy. Zhao et al. (2015) state that passive redundancy is used "when replacement of components during the operation of the system is possible". Further, they note that active redundancy is used when "replacement of components during the operation of the system is impossible". In Figure 4.1 an example model of active and passive redundancy is presented (Macedo and Guedes, 2014). While this example may appear technological, in principle it is applicable to any system.

Passive redundancy works in such a way that the primary component (C^1) is functioning at the time t=0. If C^1 fails, C^2 (the spare component) will be

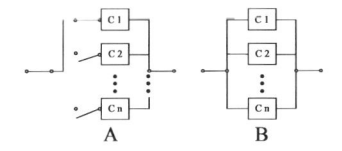

Figure 4.1 Passive (A) and active (B) redundancy
Source: Macedo and Guedes, 2014

activated. The same follows if C^2 fails, C^3 will be activated. When the nth spare component fails, the system will fail. For the active redundancy model, the spare components are operating in parallel with the primary component. When the failure occurs, the workload is automatically shared with one or more of the spare components.

It is also of interest to ascertain whether the redundancy should be at the component or the system level. Experts have long disagreed on this point. Design engineers commonly held that redundancy at the component level is always better than redundancy at the system level (Barlow and Proschlan, 1981). Boland and El-Neweihi (1995) refine the case by concluding that redundancy at the component level is better than redundancy at the system level for series systems, while the opposite holds true for parallel systems. However, this has not been properly researched in the 'system of systems' era, namely, taking into account the current interdependencies, and simulated tests produce non-ambiguous recommendations (cf. Zhao et al., 2015).

The upshot of this engineering-dominated (and still useful) debate is that any crisis manager should consider the critical functions of their system or organisation, with the aim of finding alternative solutions in the event of an unexpected occurrence that would paralyse the system or one of its parts.

4.5 Agreements and pre-arrangements with external actors

In many fields, when a crisis hits, the organisation's internal capacities and capabilities may be inadequate, or the nature of the crisis may necessitate cooperation and coordination with external actors. In terms of preparedness, this means that an organisation needs pre-arranged, coordinated and inter-operative communication, cooperation and assistance networks with the external actors concerned.

Let us take a couple of examples. The Arctic Council, an intergovernmental organisation consisting of eight member states, has several binding agreements, one of which is the Agreement on Cooperation on Marine Oil Pollution (Arctic Council, 2013). The need for this type of agreement is obvious. A large-scale accident would necessitate mobilising all available resources, and if the code of conduct and related procedures were not agreed upon and arranged in advance, a coordinated common response would be much more difficult. The agreement applies with respect to oil pollution incidents that occur in or may pose a threat to any marine area over which a signed state exercises sovereignty, sovereign rights or jurisdiction, including its internal waters, territorial sea, exclusive economic zone and continental shelf. While the preparedness would be based on combined national resources, the agreement nonetheless obliges the parties to establish, at a minimum, a national contingency plan or plans for preparedness and response to oil pollution incidents. These plans should include a minimum level of pre-positioned oil-spill combating equipment, a programme of exercises for oil pollution response organisations and training of relevant personnel, plans and communication capabilities for responding to an oil pollution incident, and a mechanism or arrangement to coordinate the response to an oil pollution incident with the capabilities to mobilise the necessary resources. The agreement also rules on incident monitoring and notification, how the request for assistance should be made, how to move resources across borders, reimbursement for costs of assistance, joint review of response operations, exchange of information, joint exercises and training activities, and so forth. The agreement proper is supplemented by more concrete and practical guidelines.

Pre-arrangements do not necessarily have to be binding. In this sense, they are only guidelines, but are still useful as a code of conduct and reference points when a crisis occurs. This is the case, for instance, with the European Union host nation support (European Commission, 2012). In many crises or disaster situations, a country has to rely on external assistance. Sometimes even civil protection teams have to be sent to another country. This would not be easy without planning and pre-arranged rules and organisation, in the event that the teams were not completely self-sufficient. The European Commission guidelines define, for instance, the procedures for requesting and offering support, with streamlined templates. The guidelines also state that the host country should be responsible for, and take appropriate measures to address, the safety and security of personnel serving in the incoming teams and modules, and of the locations, facilities, means of transport, equipment and goods used in connection with the international assistance provided. For the sake of interoperability, the host country and the incoming teams should use a common emergency communication and information system. The host country should also plan for the logistics, including route planning, the provision of necessary transport arrangements such as vehicles, escorts, maps, material handling equipment, fuel and food, as well as facilitate the use of telecommunications. These types of arrangements are also quite detailed when it comes to financial responsibilities and liability issues.

4.6 Capability-building

Capability-building refers to the skills and knowledge required for a particular task. An organisation may have the capacity to manage crises, but lack certain key capabilities. A crisis situation differs so radically from normal organisational management processes that in order to be successful in a real crisis, an organisation should be trained both in terms of individual skills and knowledge as well as organisational capabilities.

It is crucial for the capabilities to be evaluated and built in the light of anticipated threats, preferably with the help of specific scenarios (Drennan, McConnell and Stark, 2015, p. 134). Many capability-building guidelines work as checklists of sorts (e.g. Watters, 2014, pp. 178–189; Radanovsky and McDougall, 2010, pp. 110–134; APSC, 2003). They may set out principles recommending that the capability-building should be aligned with the organisation's main tasks and mission and integrated with other processes in the organisation, for example, or that one should create a capability-building culture that is recognised as legitimate and valued.

Stern (2014, pp. 6–7; cf. Haddow, Bullock and Coppola, 2011, pp. 116–123; Woodbury, 2014), referring to the US crisis management approach, has divided capability-building activities into education, training, drills and exercises. Education encompasses open-ended, long-term preparation for future endeavours. Training concerns individuals or teams developing new skills to be used in a certain position in the event of a crisis. Drills are for improving and maintaining proficiency in procedures or skills. If they test more than one function or activity, these types of activities are often called functional exercises. Exercises in turn would give highly skilled practitioners opportunities to practise together. In practice, these concepts are often blended, and their forms vary considerably.

While it is clear that any organisation should ensure that the personnel responsible for crisis management are suitably qualified in terms of education and training, possibly validated with certificates and experience, out of all the capability-building activities, exercises might be the most important ones. Exercises typically aim at improving crisis decision-making, communication and coordination procedures (e.g. Prizia, 2008). Hence, an exercise is basically a simulation of an emergency situation. Following the UK Cabinet Office guidelines (UK, 2014) – prepared for typical all-hazard emergencies but basically applicable to all kinds of exercises related to crisis management – exercises have three main purposes. First, they serve to validate the contingency plan. Second, they develop staff competencies and give them practice in carrying out the roles that they are assigned in the contingency plan. Third, exercises are also a test of how well-established the structures and procedures for crisis management are.

According to the UK guidelines, exercises can be roughly divided into four types. Discussion-based exercises are the most economical to run and easiest to prepare and are used to develop awareness about the contingency plan through discussion. Tabletop exercises are also relatively cheap and based on simulation, involving a realistic scenario and a timeline. The real time is accelerated in tabletop exercises, but can also be slowed down to consider some particular process more carefully. These kinds of exercises can be conducted in a single room, or in a series of linked rooms, which simulate the real needs of decision-making, communication and coordination. Tabletop exercises can be used to validate the contingency plan, exploring its weaknesses in procedures and finding gaps. They facilitate analysis of a crisis situation in an informal, stress-free environment. Full-scale exercises are a live rehearsal for implementing a contingency plan. These types of exercises are often the only means of really testing logistics, communications and physical resources. The problem is that they are naturally very time-consuming and resource-demanding. A fourth type of exercise would then combine certain elements of those above.

Further, the UK guidelines distinguish between exercises simulating an emergency or a crisis in a single location or in multiple locations simultaneously. In the former case, it may be a question of a fixed site (e.g. industrial plant, school, airport, train station, sports stadium, town or city centre), corridor (e.g. railway, motorway, air corridor, fuel pipeline) or an unpredictable location (e.g. bomb, chemical tanker, random shooting). In the case of multiple locations, an emergency can take place, for instance, at different sites simultaneously (e.g. explosions), within a wide or large area (e.g. toxic cloud; loss of electricity, gas, water or telephone supply; river or coastal flooding; dam or reservoir failure), or the whole national or wider international area (e.g. extreme weather; health emergencies, including influenza pandemic; foot-and-mouth disease).

Different types of seminars and discussion-based exercises should not be regarded as being of less importance than the more concrete exercises, however. As Borell and Erikson (2013) note, competencies such as coordination of action, communication and decision-making, trained through exercises, are dependent on higher-order cognitive skills and rely heavily on the conceptual capabilities of individuals, which can be advanced in discussion-based environments. Discussion-based group exercises enable one to introduce variations in the original scenario (e.g. duration of power outage), open them up to discussion and consider the conceptual and strategic differences that follow when one central parameter is changed.

Berlin and Carlström (2015) have argued that while the subject of crisis exercises is an under-researched field, the literature nonetheless attests to the fact that collaborative elements between different disciplines or types of organisations in exercises contribute positively to perceived learning; and learning, in turn, has a perceived beneficial effect on actual emergency work. Therefore, in crisis exercises where several actors or different types of organisations are involved, one should focus on this inter-organisational collaboration. Without proper training, there are special difficulties in creating cross-boundary, seamless collaboration between different organisations. Collaborating organisations want to work sequentially and in parallel rather than synchronously, often experiencing difficulty understanding the meaning of each other's behaviour, culture, rules, concepts, symbols and hierarchical levels. Each organisation tends to focus on its own tasks instead of seeing the big picture. Exercises may and should focus on filling these gaps.

Recent years have seen the expansion of different kinds of online or virtual platforms dedicated to crisis training in the form of serious games and simulation, often accompanied by detailed crisis scenarios (Tena-Chollet, 2016; Capuano and King, 2015; Nathan, 2015; Boin, 2014; Nieto-Comez, 2014; Walsh and Law, 2014; Di Loreto et al., 2013). The subjects often include such skills as anticipation, communication, teamwork, stress management, decision-making and leadership, facilitating the transposition of learned skills from theory to practice. As with any training session or exercise, crisis management also requires planning, preparation, practice and reflective debriefing. Following the current approach in learning theory, the focus should be on interactive learning and learning results, rather than on instruction-based learning.

88 *Preparedness*

Unlike in typical crisis exercises within an organisation, the virtual environment encourages interaction between learners who do not necessarily know each other. The current technology enables the use of different types of dynamics simulators and graphical user interfaces. The training session or exercise should simulate the real crisis, whereby a situation starts with important values being threatened, uncertain information and time pressures. Role plays can be applied, either for learners to deepen their normal crisis management capacity or for learning what it would be like to make decisions in some other role.

Merely practising is not enough, however. One of the basic goals of exercises is to find possible gaps or errors in the institutional structure or thinking. Exercises should lead to learning and change. However, oftentimes when exercises are conducted, this does not lead to any adjustments in policies, practices and behaviours. All too often there are barriers to change, as organisations tend to defend the status quo and play down even the identified need for change. A crisis manager may find it difficult to battle against an organisational culture, at least in organisations other than so-called high-reliability organisations where preparedness matters are taken more seriously. Too often, and too late, the need for change is acknowledged only after a crisis has taken place, creating a kind of window of opportunity to change the existing arrangements (McConnell and Drennan, 2006, pp. 67–69).

4.7 Early warning system as preparedness

Warning can be understood as a capacity that belongs to the preparedness phase of the crisis management cycle, referring to monitoring events to spot signals and indicators that predict the location, timing and magnitude of future or immediate crises (e.g. Veenema and Woolsey, 2012). In many cases, however, warning – or identification and notification of first crisis signals – is also understood as a starting point for crisis response (e.g. Watters, 2014, pp. 199–201; Harvard Business School, 2004, pp. 54–64). This is natural given that warning includes response in the sense that the response can be more or less timely. Warning should provide extra room for at least some preparatory or mitigating measures, such as evacuation or preparing the necessary equipment for an early response. Due to its important role in crisis management, the identification of a crisis and the subsequent warning are sometimes treated as a separate phase between preparedness and response (e.g. Fink, 2002, pp. 71–79).

As such, the label under which one considers crisis monitoring and warning is not the most important issue, as long as one has an efficient system in place. We will consider here the system of early warning in particular, and return to the issue in the next response-focused chapter, which discusses more dynamic issues of early warning such as why it sometimes fails. The system focus is based on the idea that the concepts of monitoring and warning should not be understood too rigidly as one-time actions. Indeed, one speaks more often in terms of early warning. If mere warning usually concentrates on alerts and warning communication, a broader definition has increasingly been adopted, focusing precisely on the concept of early warning, which starts as early as the risk assessment phase, discussed above as the first phase in the crisis management cycle.

The concept of early warning is applied in many different contexts, such as conflict management, military strategy, crime prevention, technology-related industries, information technology applications, medical care, businesses, financial systems and disaster risk and emergency management. The body of early warning literature is huge but fragmented across disciplines and application areas. Consequently, the definitions and respective applications vary largely according to the particular context in which the concept is used. Obviously, early warning practice can also vary from qualitative intelligence to quantitative measurement methodologies and tools, from indigenous traditional approaches to most developed technological solutions. It has been argued that there cannot be a general theory of how to build warning systems that does not take account of local problems. Therefore, there is no single best or generic way to build an early warning system (EWS); rather, it always depends on the context even when limited to one specific field such as national defence (Bracken, 2008). It would be much more difficult to apply the same EWS model for, say, a military strategic surprise (Bracken, Bremmer and Gordon, 2008) or a currency crisis (Sevim et al., 2014).

Let us therefore, for the sake of illustration, consider early warning in this section mainly from the perspective of one particular field, namely, disaster risk management. In the context of the European Union's Civil Protection Mechanism, for instance, early warning means the timely and effective provision of information that allows action to be taken to avoid or reduce risks and the adverse impacts of a disaster, and to facilitate preparedness for an effective response (European Council, 2007). UNISDR (2009) in turn defines early warning more in terms of a system. It is the set of capacities needed to generate and disseminate timely and meaningful warning information to enable individuals, communities and organisations threatened by a hazard to prepare and act appropriately and in sufficient time to reduce the possibility of harm or loss. In the definition's comments, it is further detailed that an early warning system comprises four key elements: knowledge of the risks; monitoring, analysis and forecasting of the hazards; communication or dissemination of alerts and warnings; and local capabilities to respond to the warnings received.

EWS of this kind necessitates integrated bodies that are responsible for scientific monitoring, data processing, event forecasting, translation of scientific information into meaningful warnings for the target audience, and the widest possible dissemination of warnings to those who could be affected by the hazards (cf. UK Parliament, 2012). Golnaraghi (2012) in her edited book reviewing several national disaster risk management systems has summarised the guidelines that should characterise a good EWS into ten main points. In a way they attest to the fact that an EWS should not be isolated from the general crisis management system, but rather form an integral part, a message that is probably applicable to EWS in any field. First, in order to be effective, there has to be strong political recognition of the benefits of EWS, reflected in harmonised national to local disaster risk management policies, planning, legislation and budgeting. Second, an effective EWS should be built upon four components: hazard detection, monitoring and forecasting; analysing risks and incorporation of risk information in emergency planning and warnings; disseminating timely and authoritative warnings, as well as community planning and preparedness and the ability to activate emergency plans to prepare and respond, with coordination across agencies involved in EWS, at national to local levels. Third, EWS stakeholders should be identified and their roles, responsibilities and coordination mechanisms clearly defined and documented within national to local plans, legislation and directives. Fourth, EWS capacities need to be supported by adequate resources (human, financial, equipment, etc.) across national to local levels and the system designed and implemented accounting for long-term sustainability factors. Fifth, hazard, exposure and vulnerability information should be used to carry out risk assessments at different levels, as critical input into emergency planning and development of warning messages. Sixth, warning messages should be clear, consistent and include risk information, as well as designed with due consideration of linking threat levels to emergency preparedness and response actions (by using colour, flags, etc.) understood by the authorities and the population, and issued from a single, recognised and authoritative source. Seventh, warning dissemination mechanisms must be able to reach the authorities, other EWS stakeholders and the population at risk in a timely and reliable fashion. Eighth, emergency response plans should be developed taking into account hazard or risk levels, characteristics of the exposed communities, coordination mechanisms and various EWS stakeholders. Ninth, training on risk awareness, hazard recognition and related emergency response actions should be integrated into various formal and informal educational programmes and linked to regularly conducted drills and tests across the system to ensure operational readiness at any time. Finally, effective feedback and improvement mechanisms need to be in place at all levels of EWS to provide systematic evaluation and ensure system improvement over time.

References

Alexander, D. (2005) Towards the Development of a Standard in Emergency Planning. *Disaster Prevention and Management: An International Journal*, 14(2), pp. 158–175.

Allison, G.T. (1971) *Essence of Decision: Explaining the Cuban Missile Crisis*. Boston: Little, Brown and Company.

Allison, G.T. and Zelikow, P. (1999) *Essence of Decision: Explaining the Cuban Missile Crisis*. Second edn. New York: Longman.

Ansell, C., Boin, A. and Keller, A. (2010) Managing Transboundary Crises: Identifying the Building Blocks of an Effective Response System. *Journal of Contingencies and Crisis Management*, 18(4), pp. 195–207.

APSC. (2003) Building Capability: A Framework for Managing Learning and Development in the APS. Australian Public Service Commission. Available at: http://www.apsc.gov.au/__data/assets/pdf_file/0008/50975/capability.pdf

ARCTIC COUNCIL. (2013) Agreement on Cooperation on Marine Oil Pollution Preparedness and Response in the Arctic. Available at: https://oaarchive.arctic-council.org/bitstream/handle/11374/529/EDOCS-2067-v1-ACMMSE08_KIRUNA_2013_agreement_on_oil_pollution_preparedness_and_response__in_the_arctic_formatted.PDF?sequence=5&isAllowed=y

Baker, G.H. (2005) A Vulnerability Assessment Methodology for Critical Infrastructure Facilities. Available at: http://www.jmu.edu/iiia/wm_library/Vulnerability_Facility_Assessment_05-07.pdf

Barlow, R. and Proschan, R. (1981) *Statistical Theory of Reliability and Life Testing*. Silver Spring, MD: Madison.

Bendor, J. (2010a) *Bounded Rationality and Politics*. Los Angeles: University of California Press.

Bendor, J. (2010b) Institutions and Individuals. In Bendor, J., *Bounded Rationality and Politics*. Los Angeles: University of California Press, pp. 63–182.

Bendor, J. and Hammond, T.H. (1992) Rethinking Allison's Models. *The American Political Science Review*, 86(2), pp. 301–322.

Berlin, J.M. and Carlström, E.D. (2015) Three-Level Collaboration Exercise – Impact of Learning and Usefulness. *Journal of Contingencies and Crisis Management*, 23(4), pp. 257–265.

Boin, A. (2014) *European Platforms for Crisis Management Training: CRIMSON, INDIGO and VASCO*. In Stern (2014), op. cit., pp. 51–53.

Boin, A. and 't Hart, P. (2003) Public Leadership in Times of Crisis: Mission Impossible? *Public Administration Review*, 63(5), pp. 544–553.

Boin, A. and Lagadec, P. (2000) Preparing for the Future: Critical Challenges in Crisis Management. *Journal of Contingencies and Crisis Management*, 8(4), pp. 185–191.

Boland, P.J. and El-Neweihi, E. (1995) Component Redundancy Versus System Redundancy in the Hazard Rate Ordering. *IEEE Transaction on Reliability*, 44, pp. 614–619, 1005.

Borell, J. and Erikson, K. (2013) Learning Effectiveness of Discussion-based Crisis Management Exercises. *International Journal of Disaster Risk Reduction*, 5, pp. 28–37.

Bracken, P. (2008) How to Build a Warning System. In Bracken, P., Bremmer, I. and Gordon, D. (eds) *Managing Strategic Surprise: Lessons from Risk Management and Risk Assessment*. Cambridge: Cambridge University Press, pp. 16–41.

Burlin, W.K. and Hyle, A.E. (1997) Disaster Preparedness Planning: Policy and Leadership Issues. *Disaster Prevention and Management: An International Journal*, 6(4), pp. 234–244.

Capuano, N. and King, R. (2015) Knowledge-Based Assessment in Serious Games: An Experience on Emergency Training. *Journal of E-Learning and Knowledge Society*, 11(3), pp. 117–132.

Carmeli, A. and Schaubroeck, J. (2008) Organisational Crisis-Preparedness: The Importance of Learning from Failures. *Long Range Planning*, 41, pp. 177–196.

Di Loreto, I. *et al.* (2013) Supporting Crisis Training with a Mobile Game System. In Minhua, M., Oliveira, F., Petersen, S., and Baalsrud, H.J. (eds) *Serious Games Development and Applications*. 4th International Conference, SGDA, Trondheim, Norway, September 25–27. Springer, pp. 165–177.

Drennan, L., McConnel, A. and Stark, A. (2015) *Risk and Crisis Management in the Public Sector*. Second edn. New York: Routledge.

DSB. (2016) Kommuneundersøkelsen 2016. Status for samfunnssikkerhets- og beredskapsarbeidet i kommunene. *Direktoratet for samfunnssikkerhet og beredskap*. Available

at: https://www.dsb.no/globalassets/dokumenter/rapporter/kommuneundersokelsen_2016pdf
Dynes, R., Quarantelli, E.L. and Kreps, G. (1972) *A Perspective on Disaster Planning*. Columbus, OH: Ohio State University Disaster Research Center.
Egan, M. (2007) Anticipating Future Vulnerability: Defining Characteristics of Increasingly Critical Infrastructure-like Systems. *Contingencies and Crisis Management*, 15(1), pp. 4–17.
Emmens, B. et al. (2008) Understanding Surge Capacity within International Agencies. Available at: http://odihpn.org/magazine/understanding-surge-capacity-within-international-agencies/
Enander, A., Hede, S. and Lajksjö, Ö (2015) Why Worry? Motivation for Crisis Preparedness Work among Municipal Leaders in Sweden. *Journal of Contingencies and Crisis Management*, 23(1), pp. 1–10.
European Commission. (2012) *EU Host Nation Support Guidelines*. Commission Staff Working Document. Brussels, SWD(2012) 169 final.
European Council. (2008) Council Directive 2008/114/EC of 8 December 2008 on the identification and designation of European critical infrastructures and the assessment of the need to improve their protection (Text with EEA relevance). *Official Journal of the European Union*, L 345, pp. 75–82.
European Council. (2013) Decision No 1313/2013/EU of the European Parliament and of the Council of 17 December 2013 on a Union Civil Protection Mechanism (Text with EEA relevance). *Official Journal of the European Union*, L 347, pp. 924–947.
Fiksel, J. (2003) Designing Resilient, Sustainable Systems. *Environmental Science and Technology*, 37, pp. 5330–5339.
Fink, S. (2002) *Crisis Management: Planning for the Inevitable*. Lincoln, NE: iUniverse Inc.
Fujin, C. and Tiehan, T. (eds) (2008) *Public Crisis Management in China*. Beijing: Foreign Languages Press.
Golnaraghi, M. (2012) An Overview: Building a Global Knowledge Base of Lessons Learned from Good Practices in Multi-Hazard Early Warning Systems. In Golnaraghi, M. (ed.) *Institutional Partnerships in Multi-Hazard Early Warning Systems: A Compilation of Seven National Good Practices and Guiding Principles*, Heidelberg: Springer-Verlag, pp. 1–8.
Greighton, P.V. (ed.) (2009) *Emergency Preparedness*. New York: Nova.
Groenendaal, J., Helsloot, I. and Scholtens, A. (2013) A Critical Examination of the Assumptions Regarding Centralized Coordination in Large-Scale Emergency Situations. *Homeland Security and Emergency Management*, 10(1), pp. 113–135.
Haddow, G.D., Bullock, J.A. and Coppola, D.P. (2011) *Introduction to Emergency Management*. Fifth edn. Amsterdam: Elsevier.
Harvard Business School. (2004) *Crisis Management: Master the Skills to Prevent Disasters*. Boston, MA: Harvard Business School Press.
Heath, R. (1998) Dealing with the Complete Crisis: The Crisis Management Shell Structure. *Safety Science*, 30, pp. 139–150.
Hémond, Y. and Robert, B. (2012) Preparedness: The State of the Art and Future Prospects. *Disaster Prevention and Management: An International Journal*, 21(4), pp. 404–417.
IFRC. (2012) Contingency Planning Guide. Geneva: International Federation of Red Cross and Red Crescent Societies.
ISO/PAS. (2007) Societal Security – Guidelines for Incident Preparedness and Operational Continuity Management. ISO/PAS 22399.

ISO. (2007) Security Management Systems for the Supply Chain. Guidelines for the Implementation of ISO 28000. 28004.
ISO. (2009) Risk Management – Principles and Guidelines. ISO 31000.
ISO. (2011) Security Management Systems for the Supply Chain – Development of Resilience in the Supply Chain. 28002.
ISO. (2012) Societal Security – Business Continuity Management Systems – Requirements. ISO 22301.
ISO. (2014a) Security Management Systems for the Supply Chain. Guidelines for the Implementation of ISO 28000. Part 2: Guidelines for Adopting ISO 28000 for Use in Medium and Small Seaport Operations. 28004-28002.
ISO. (2014b) Security Management Systems for the Supply Chain. Guidelines for the Implementation of ISO 28000. Part 3: Additional Specific Guidelines for Adopting ISO 28000 for use of Medium and Small Businesses (Other than Marine Ports). 28004-28003.
ISO. (2014c) Security Management Systems for the Supply Chain. Guidelines for the Implementation of ISO 28000. Part 4: Additional Specific Guidelines for Adopting ISO 28000 if Compliance with ISO 280001 Is a Management Objective. 28004-28004.
ISO. (2015) Environmental Management Systems – Requirements with Guidance for Use. ISO 14001.
Johansen, W., Aggerholm, H. and Frandsen, F. (2012) Entering New Territory: A Study of Internal Crisis Management and Crisis Communication in Organizations. *Public Relations Review*, 38(2), pp. 270–279.
Kaneberg, E., Hertz, S. and Jensen, L.-M. (2016) Emergency Preparedness Planning in Developed Countries: The Swedish Case. *Journal of Humanitarian Logistics and Supply Chain Management*, 6(2), pp. 145–172.
Kuipers, S. et al. (2015) Building Joint Crisis Management Capacity? Comparing Civil Security Systems in 22 European Countries. *Risk, Hazards & Crisis in Public Policy*, 6(1), pp. 1–21.
Labaka, L, Hernantes, J. and Sarriegi, J.M. (2015) Resilience Framework for Critical Infrastructure: An Empirical Study in a Nuclear Plant. *Reliability Engineering and System Safety*, 141, pp. 92–105.
Landstedt, J. and Holmström, P. (2007) Electric Power Systems Blackouts and the Rescue Services: The Case of Finland. Stockholm: CIVPRO Working Paper.
Lindblom, C.E. (1959) The Science of Muddling Through. *Public Administration Review*, 19, pp. 79–88.
Lunde, I.K. (2014) *Praktisk krise- og beredskapsledelse*. Oslo: Universitetsforlaget AS.
Macedo, D. and Guedes, L.A. (2014) A Dependability Evaluation of Internet of Things Incorporating Redundancy Aspects. In *Networking, Sensing and Control (ICNSC)*, 2014 IEEE 11th International Conference, pp. 417–422.
McConnell, A. and Drennan, L. (2006) Mission Impossible? Planning and Preparing for Crisis. *Journal of Contingencies and Crisis Management*, 14(2), pp. 59–70.
Menzies, P. (2014) Counterfactual Theories of Causation. In Zalta, E.N. (ed.) *The Stanford Encyclopedia of Philosophy (Spring 2014 Edition)*. Available at: http://plato.stanford.edu/archives/spr2014/entries/causation-counterfactual/
MSB. (2016) Nationell beredskapsplan för hanteringen av en kärnteknisk olycka. Available at: https://www.msb.se/Upload/Insats_och_beredskap/CBRN/150130%20Beredskapsplan%20f%C3%B6r%20hantering%20av%20k%C3%A4rnteknisk%20olycka.pdf

NASA. (2016) Active Redundancy. Practice No. PD-ED-1216. Available at: https://oce.jpl.nasa.gov/practices/1216.pdf

Nathan, M.L. (2015) e-Learning's Indispensability in Crisis Management Education. *Global Education Journal*, 3, pp. 82–88.

Nieto-Comez, R. (2014) *STANCE: An Asynchronic Training and Evaluation Tool for Strategic Leadership*. In Stern (2014), op. cit., pp. 53–58.

NIST. (2010) *Contingency Planning Guide for Federal Information Systems*. NIST Special Publication 800-834 Rev. 1.

Parnell, J.A. (2015) Crisis Management and Strategic Orientation in Small and Medium-Sized Enterprises (SMEs) in Peru, Mexico and the United States. *Journal of Contingences and Crisis Management*, 23(4), pp. 221–233.

Paton, D. (2003) Disaster Preparedness: A Social-Cognitive Perspective. *Disaster Prevention and Management: An International Journal*, 12(3), pp. 210–216.

PBS. (2011) PBS I. Politiets BEREDSKAPSSYSTEM del I. Retningslinjer for politiets beredskap. Norway: POD Available at: https://www.politi.no/vedlegg/rapport/Vedlegg_1690.pdf

Pearson, C.M. and Mitroff, C.M. (1993) From Crisis Prone to Crisis Prepared: A Framework for Crisis Management. *The Executive*, 7(1), pp. 48–59.

Perry, R.W. and Lindell, M.K. (2003) Preparedness for Emergency Response: Guidelines for the Emergency Planning Process. *Disasters*, 27(4), pp. 336–350.

Pinkowski, J. (2008) Coastal Development and Disaster Preparedness: The Delusion of Preparedness in Face of Overwhelming Forces. In Pinkowski, J. (ed.), *Disaster Management Handbook*. Boca Raton, FL: CRC Press, pp. 3–17.

Piotrowski, C. (2006) Hurricane Katrina and Organization Development: Part 1. Implications of Chaos Theory. *Organization Development*, 24(3), pp. 10–19.

Power, R. (2008) Issues in Hospital Preparedness. In Pinkowski, J. (ed.), *Disaster Management Handbook*. Boca Raton, FL: CRC Press, pp. 561–570.

Prizia, R. (2008) The Role of Training in Disaster Management: The Case of Hawaii. In Pinkowski, J. (ed.), *Disaster Management Handbook*. Boca Raton, FL: CRC Press, pp. 529–552.

Pursiainen, C. and Gattinesi, P. (2014). *Towards Testing Critical Infrastructure Resilience*. Publications Office of the European Union, JRC Scientific and Policy Reports.

Pursiainen, C., Hedin, S. and Hellenberg, T. (2005) Civil Protection Systems in the Baltic Sea Region. Towards Integration in Civil Protection Training. *Eurobaltic Publications*, 3. Helsinki: Aleksanteri Institute.

Quarantelli, E.L. (1984) Organizational Behavior in Disasters and Implications for Disaster Planning. In *Monographs of the National Emergency Training Center*, 1(2), pp. 1–31.

Radvanovsky, R. and McDougall, A. (2010) *Critical Infrastructure. Homeland Security and Emergency Preparedness*. Second edn. Boca Raton, FL: CRC Press.

Sager, T. (2013) *Reviving Critical Planning Theory: Dealing with Pressure, Neo-Liberalism, and Responsibility in Communicative Planning*. London: Routledge.

Scholtens, A., Jorritsma, J. and Helsloot, I. (2014) On the Need for a Paradigm Shift in the Dutch Command and Information System for the Acute Phase of Disasters. *Journal of Contingencies and Crisis Management*, 22(1), pp. 39–51.

Schwarzer, D. (2015) Building the Euro Area's Debt Crisis Management Capacity with the IMF. *Review of International Political Economy*, 22(3), pp. 599–625.

Sevim, C. et al. (2014) Developing an Early Warning System to Predict Currency Crises. *European Journal of Operational Research*, 237(3), pp. 1095–1104.

Simon, H.A. (1972a) *Administrative Behavior*. New York: Free Press.

Simon, H.A. (1972b) Theories of Bounded Rationality. In McGuire, C.B. and Radner, R. (eds.), *Decision and Organization*. Amsterdam: New-Holland Publishing Company, pp. 361–377.

Simpson, D.M. (2008) Disaster Preparedness Measures: A Test Case Development and Application. *Disaster Prevention and Management: An International Journal*, 17(5), pp. 645–661.

Stern, E. (2013) Preparing: The Sixth Task of Crisis Leadership. *Journal of Leadership Studies*, 7(3), pp. 51–56.

Stern, E. (ed.) (2014) *Designing Crisis Management Training and Exercises for Strategic Leaders: A Swedish and United States Collaborative Project*. Stockholm: Swedish National Defense College.

Sturgis, R. (2008) Strategic Planning for Emergency Managers. In Pinkowski, J. (ed.), *Disaster Management Handbook*. Boca Raton, FL: CRC Press, pp. 571–581.

Tammepuu, A., Tammepuu, O. and Sepp, K. (2009) Emergency Preparedness in Integrated Management Systems: Case Study of the Port of Tallinn. In Duncan, K. and Brebbia, C. (eds) *Disaster Management and Human Health Risk: Reducing Risk, Improving Outcomes*. Boston: WIT Press, pp. 65–76.

Tena-Chollet, F. et al. (2016) Training Decision-Makers: Existing Strategies for Natural and Technological Crisis Management and Specifications of an Improved Simulation-Based Tool. *Safety Science*. Available at: http://www.sciencedirect.com/science/article/pii/S092575351630039X

UK Cabinet Office. (n.d.) Section A: Introduction, Definitions and Principles of Infrastructure Resilience. Available at: https://www.gov.uk/government/uploads/system/uploads/attachment_data/file/78902/section-a-natural-hazards-infrastructure.pdf

UK Parliament. (2012) Early Warning for Natural Disasters. In Ghosh, S. (ed.), *Natural Disaster Management. New Technologies and Opportunities*. India: The Icfai University Press, pp. 15–24.

UK Parliament. (2014) Public Safety and Emergencies – Guidance. Emergency Planning and Preparedness: Exercises and Training. Government of the United Kingdom. Cabinet Office and National Security and Intelligence. Available at: https://www.gov.uk/guidance/emergency-planning-and-preparedness-exercises-and-training

UNISDR. (2009) UNISDR Terminology on Disaster Risk Reduction, United Nations International Strategy for Disaster Reduction (UNISDR), Geneva, Switzerland. Available at: http://www.unisdr.org/we/inform/terminology

Veenema, T.G. and Woolsey, C. (2012) Essentials of Disaster Planning. In Veenema, T.G. (ed.) *Disaster Nursing and Emergency Preparedness: for Chemical, Biological, and Radiological Terrorism and Other Hazards*. Third edn. New York: Springer Publishing Company, pp. 1–20.

Walsh, W. and Law, R. (2014) *Virtual Games as a Key to Advancing Strategic Leadership Education*. In Stern, E. (2014), op. cit., pp. 42–50.

Watters, J. (2014) *Disaster Recovery, Crisis Response & Business Continuity: A Management Desk Reference*. California: Apress.

Woodbury, G. (2014) *NPGS CHDS Executive Education Seminar (Mobile Education Team Program)*. In Stern, E. (2014), op. cit., pp. 27–39.

Zhao, P. et al. (2015) Redundancy Allocation at Component Level Versus System Level. *European Journal of Operation Research*, 241, pp. 402–411.

5 Response

Crises demand some form of response because passivity can be costly. The response naturally varies considerably depending on the kind of crisis in question. From the perspective of large-scale societal disasters, no matter what the root cause is, UNISDR (2009) defines a response as "The provision of emergency services and public assistance during or immediately after a disaster in order to save lives, reduce health impacts, ensure public safety and meet the basic subsistence needs of the people affected." Mention is also made of the fact that disaster response is predominantly focused on "immediate and short-term needs" and is sometimes called disaster relief.

If we were to focus on political crises or business reputation crises instead, we might not even use the concept of response. Terms such as crisis containment, crisis resolution, damage limitation or simply crisis decision-making might be more appropriate. In any event, the issue concerns what to do when the crisis is imminent, while it is ongoing, or immediately after it has occurred.

This chapter discusses what response means, in both theoretical and practical-normative terms. In the following sections, some of the most central themes in the respective literature are outlined. This discussion is divided into four somewhat overlapping sections. First, the very construction of a crisis is considered – how it is identified, framed and made sense of. Second, the chapter deals with the rather practical issue of how to organise crisis response, moving, however, on a rather generic level. The third section is the most theoretical one. It reviews the main schools of thought that are applied in an attempt to understand the nature of crisis decision-making, highlighting the way in which it differs from decision-making in normal situations. This type of literature is usually the basis when trying to explain crisis decision-making post factum. Finally, the chapter provides a short overview of the relatively under-theorised theme of crisis communication, discussing its various dimensions. Through these four main categories and their sub-themes, that are common to all kinds of crises, the chapter covers most of the issues of relevance to the response phase of a crisis.

5.1 Identifying and framing a crisis

Identifying a crisis might be obvious in many cases, but more ambiguous and contested in others. So-called creeping crises are particularly difficult to identify as such. Sometimes threats to core values, time pressures and uncertainty, even if objective to some extent, do not produce any sense of crisis. By contrast, a situation is sometimes perceived as a crisis when the real threat to basic values is actually minimal. Whether, and to what extent, a situation is seen as a crisis depend on the beliefs and perceptions of the decision-makers, in other words, on problem-framing or problem representation (Sylvan, 1998). This brings us back to the human factors discussed earlier in Chapter 3. Obviously, the objective properties of the situation matter to a degree – in the case of, say, an earthquake or a terrorist attack – but few situations are self-evidently crises. On the contrary, they can be defined as normal policy processes, routine rather than crisis situations, or as non-events if signals are lost because of the surrounding noise or if they are by purpose omitted.

Independently of their subjective perception of the situation, decision-makers still have the option of publicly calling it a crisis. On the one hand, decision-makers can deliberately define a situation as a crisis even if they do not perceive that it entails any major danger or time pressure. They may do so in order to force through policies and actions with extraordinary powers and a sense of urgency that they would not acquire otherwise ('t Hart, 1993; Boin, 't Hart and McConnell, 2009). In this case, the audience – the media, the political opposition and the public – needs to accept the definition of the situation as a crisis. On the other hand, decision-makers are sometimes unwilling to define a situation as a crisis and prefer to regard the event as a routine matter. In fact, because of the inherent negative connotations of the term 'crisis', decision-makers often opt for other terms when describing a situation that fulfils the definitional characteristics of a crisis.

Yet the media, political opposition or the general public often force leaders to acknowledge that the situation is indeed a crisis. The media, in particular, can create a crisis by exaggerating the danger of the situation, thereby threatening at least one core value of the decision-makers, namely, credibility and survival. The media can also exert considerable time pressure on the decision-makers to do something about the situation (Robertson, 2001; Forsberg and Pursiainen, 2006). A recurrent observation is that the initial problem-framing can prove to be influential in setting the tone for the subsequent crisis management process.

A certain situation is often recognised as a crisis on the basis of similar historical cases, and thus the lessons learned from history also guide the decision-making. In Hirschbein's (1997, p. 3) words, a "situation is constructed as a crisis when it is likened to a previous crisis narrative indelibly etched in an actor's memory and inscribed in his texts – a saga remembered as a critical juncture requiring urgent perilous choices". It is notable, however, that historical analogies are used for different purposes, including learning, persuasion, manipulation, rhetoric and prediction. Moreover, decision-makers use the past either politically (purposefully) or cognitively (Brändström, Bynander and 't Hart, 2004). The avowedly political use means that a decision-maker intentionally chooses an event from the past to describe the current or forthcoming event. Sometimes, this kind of usage can be helpful in explaining a complex situation to the general public, but often this is done simply to advance a certain political line or project. Depending on which historical analogy is chosen, we can stimulate different policy approaches towards the same issue.

The cognitive use of historical analogies is less intentional. In acute crisis situations in particular, such cognitive short-cuts help one to quickly make sense of confusing events, offering a diagnosis on the basis of a past event that is perceived as sharing similar features (cf. Parker and Stern, 2002, p. 605ff). It can help us to analyse the situation quickly and to anticipate the possible consequences. However, if a faulty historical analogy is used as a cognitive shortcut to define the current case, it will either misdirect or at least narrow and constrain the analysis of the situation.

In any event, the way that a crisis is labelled and framed is thus crucial, presenting a simplified interpretation. Those involved in a crisis often try to frame it in ways that correspond with their needs and interests. This framing can involve the conscious or unconscious selective exploitation of data to advance a specific interpretation, or it can involve the issue of who is to blame (Drennan, McConnell and Stark, 2015, pp. 176, 177). Boin, 't Hart and McConnell (2009) speak about a framing contest between actors. In the first contest, the significance of the crisis can be minimised, acknowledged or maximised, while in the second contest its causes can be framed differently. In the case of a national or international crisis, a key issue might be whether or not a situation is characterised as a security crisis. According to the securitisation thesis of the so-called Copenhagen School, in labelling something as a security issue the actor has claimed the right to handle it with extraordinary means and break the normal rules of politics (Wæver, 1995). This implies that there is a tendency in security crises to shift from civilian models towards military models of dealing with the issue. It also leads to questions of when to securitise an issue and when not to, or whether decision-makers typically favour securitisation over desecuritisation.

The same issues were already discussed in part much earlier by Weick (1988; 1995), who coined the term sensemaking in a crisis situation. Weick argued, in much the same vein as the discussion above, that crises are often constructed and put into place by human actors. The point of the term sensemaking is that it focuses attention on interference due to human factors when endeavouring to understand such crises, which were previously often understood as unambiguous. He pointed to the dilemma that emerges when sorting out a crisis as it unfolds, as this often requires action, which simultaneously generates the raw material that is used for sensemaking and which affects the unfolding crisis itself. When a crisis occurs, the crisis manager cannot know in advance what he or she will be up against, but this involvement in the crisis and the actions taken, based on certain presumptions or preconceptions of what is going on, have an impact on the very crisis event itself. Thus, according to Weick, it would be wrong to say that the situation will determine the appropriate action, as the actions of the crisis manager determine the situation. In short, action affects events and can make things worse, or at least human action can amplify small deviations into major crises, and some human actions, aimed at countering the crisis, may actually set the crisis in motion. As the human responses stimulate further action, the actions themselves become increasingly important components of the crisis. This can easily be unintentional. In particular, once crisis managers become committed to an action, and then build an explanation that justifies that action, the explanation is transformed into an assumption that is taken for granted, and cannot easily be viewed as a potential contributor to a crisis. The available capacity and response repertoire also affect crisis perception, because crisis managers tend to see only those events they feel they have the capacity to do something about. Weick's basic solution for avoiding biases in sensemaking, and at least partially enhancing the quality of decision-making, is that if crisis managers are aware of the possible impact of their actions and capacities to influence the essence of the crisis, this heightened awareness could expand their view of a developing crisis and prompt them to seek alternative avenues for intervention.

Boin et al. (2005) relate sensemaking more to the early warning situation than to the actual crisis decision-making process, proposing that there should be a proper early warning system in place and that a crisis manager (or the leader of the organisation), together with the relevant staff, should routinely scan potential crisis issues, trying not to withhold bad news. Moreover, one should take nothing at face value, proactively seeking a diversity of information instead. Thus, while sensemaking has become one of the major themes in the literature dealing with crisis management since Weick's classic texts, it has become somewhat fragmented into several approaches. The differences may be reflected in questions on what sensemaking actually is, how and when it takes place, how shared it is among the crisis managers in a crisis situation, and whether it is an individual and cognitive process or more rooted in social and discursive structures. Apart from some rather recent reviews (Maitlis and Sonenshein, 2010; Maitlis and Christianson, 2014), these differences have remained rather poorly classified and articulated.

In any case, from a normative perspective, a crisis manager should always consider their own presumptions and actions critically when a crisis occurs, in order to avoid the above-discussed biases. Forsberg and Pursiainen (2006) have proposed a set of issues that could be regarded as a checklist to start from, both for scholars and decision-makers. It includes such questions as how the crisis in question arose, how it was constructed and what the crucial event was that changed the normality into a crisis. Were any ominous signals sent and detected, and an alarm sounded? Was there an existing institutional or organisational early-warning framework for identification of the crisis?

100 *Response*

One should also ask who was the first to define the situation as a crisis. What was the role of the institutions in constructing or identifying the situation as a crisis? Where did the first signals come from, and how were these signals interpreted? The crisis manager should consider whether there was a historical analogue that helped to identify the crisis, and critically scrutinise whether the analogy was lacking any idiosyncrasies pertinent to the new situation. One could also ask whether any cases of a similar kind could be construed as non-crises. Furthermore, one should always consider whether there were any contradictory interpretations of the situation of relevance to the occurrence of the crisis decision-making situation. With reference to the definition of a crisis used in the present book, one should also examine the underlying assumptions that were embedded in the debates surrounding the crisis, as well as the essential goals or important values that were under threat. One should likewise elaborate on the factors that determined the time limits and deadlines, or otherwise caused stress in decision-making, as well as those factors from which the unpredictability or uncertainty arose. In post-factum analysis, one could consider such issues as the turning points or main stages of the crisis, and how and why the crisis was defined as terminated.

5.2 Organising response

There is a body of literature that could be called practical-normative in terms of its rather low level of theoretical abstraction. That is the literature on how to organise crisis response or crisis decision-making to enhance its quality. This is an issue that already was rather exhaustively discussed in the context of preparedness, as preparedness plans usually propose ideal models to this effect.

To add to that discussion, Scholtens, Jorritsma and Helsloot (2014), for instance, in criticising the traditional control and command model, have proposed five normative criteria for strategic-level crisis decision-making. The first task of the strategic crisis leadership is the rapid dissemination of factual information to the public, within one or two hours at most. Second, the crisis manager has to be able to produce fast meaning-making. Indeed, the role of the top manager of an organisation is not so much a commander-in-chief, but a meaning-maker. Third, operational decision-making should be decentralised in order to be fast enough. Staff should not be kept waiting for orders to trickle down the hierarchy. Fourth, this means that monodisciplinary command should be favoured over multidisciplinary coordination because it saves time. Operational crisis managers should be able to rely on their own judgement when it comes to deciding which tasks should be prioritised. And finally, strategic decisions must be taken almost instantly. Post-crisis evaluations show that during the acute phase of sudden unanticipated disasters, strategic decisions are often not taken simply because uncertainty at the higher command levels causes leaders to spend so much time searching for more information that by the time they find it, the original operational situation no longer exists.

Other practical guidebooks add various issues, such as the fact that improvisation matters. Drennan, McConnell and Stark (2015, pp. 183, 184) note that there are immense pressures to stick to prior agreements detailed in the preparedness or contingency plans, because individuals may feel that they will otherwise impair the crisis response, or fear that they will face repercussions. But they emphasise the importance of situation assessment, and the possibility of unknowns. The choice of response strategy then depends on at least two factors, namely, the level of risk and the amount of available time. In a situation of low risk and plenty of time, one may allow oneself the luxury of making an analytical comparison between several pre-defined options, or even of being creative, designing a new option instead of those planned. The less time one has and the higher the risk, the better it is to rely on pre-determined rule-based procedures or patterns of response options. Inactivity may sometimes be the best strategy.

One of the issues that become important especially during an acute crisis situation is leadership. Crisis requires a leader to make decisions and an organisation to implement them. The issue of establishing the response structure and, most of all, a key decision-making unit, was already discussed in some detail in the context of preparedness. If an organisation has invested in preparedness planning, it usually has some kind of a plan for setting up the key decision-making group and command hierarchy. This plan may even have been tested and rehearsed.

However, the issue often changes when a real crisis hits. Indeed, while most organisations may have an institutional solution for a crisis situation, case studies (e.g. Forsberg and Pursiainen, 2006) have proved that when a crisis hits, the decision-making group is often formed on an ad hoc basis, even against the plan made in advance. This feature is most visible in high-level national crises. The situational dynamics and idiosyncrasies of the crisis may lead to the conclusion that the government and its various sub-committees do not necessarily provide the best possible compositions for crisis decision-making. While according to the plan it would, for instance, be appropriate to rely on the same decision-makers as in times of normality, in a real crisis situation the key decision-makers may not trust all members of the cabinet, or they may require more expert knowledge than the ministers can provide.

It is indeed difficult to create crisis management structures without knowing the exact nature of the crisis. With this in mind, Fink (2002, p. 57) has noted that there should be "a central core to each crisis management team, but a technical crisis requires technical people; a financial crisis requires financial people". Stern (2013) argues in a similar spirit that there is no single optimal form of organising for crisis management. Rather, a crisis organisation should be designed taking into account the characteristics and context of the given setting. Thus, in a crisis situation, the group or situational dynamics may often lead to improvised and spontaneous solutions in forming the decision-making unit, which contradicts both the normal-time decision-making practices as well as those designed for a crisis situation. For their part, Drennan,

McConnell and Stark (2015, pp. 161–163) have argued that when the pre-designed model seems to be failing, a new model should be – and has been in many cases – adopted even in the midst of a crisis.

As already discussed in the context of framing, given the ambiguity of a crisis situation, it provides an opportunity to change the prevailing power relations and may also be used to promote a certain political or societal enterprise. This phenomenon can be called managing by crisis rather than crisis management. One traditional key assumption about crisis decision-making, however, is that it differs from normal decision-making because it tends to become centralised, even if the crisis preparedness planning may have been based on a decentralised model. For political scientists, centralisation usually means that the size of the decision-making group becomes smaller and more closed (Forsberg and Pursiainen, 2006). For a business adviser, the normative maxim might go something like this: "Unlike normal business operations, when in crisis mode, there should be a more rigid command-and-control structure, with short lines of communication, near real-time flow of information, and rapid decision-making" (Watters, 2014, p. 202).

Yet bypassing the existing institutions can, in itself, increase distrust among the key players and contribute to the crisis negatively. The older research often tended to claim that the empirical record concerning the significance of the institutional setting remained mixed (Brecher, 1993; Haney, 1997). On the other hand, more contemporary research claims that the more complex a crisis, the more one needs adaptive self-organising networks, as they constitute the most effective approach to crisis management (Ansell, Boin and Keller, 2010). Some time ago, Rosenthal, 't Hart and Kouzmin (1991) challenged the centralisation thesis, and emphasised the effects of bureaucratic politics and struggles, whereas Kouzmin and Jarman (2004) downplayed the whole issue and argued that in crisis situations the quality of the leadership is much more important than organisational structures.

In any event, a crisis calls for leadership, and leadership requires well-known faces and clear responsibilities. This thesis can be qualified depending on the type of crisis and society in question ('t Hart et al., 1993), but it seems to hold true as a general tendency. In political decision-making studies (Breuning, 2007; Preston and 't Hart, 2011, p. 245), it is noteworthy that the forming of the decision-making unit is greatly dependent on the leader's personality and political style. Some leaders prefer more hierarchical command chains. Each department or service produces information from its respective field and delivers it to the key decision-maker. The latter then forms an overall picture of this material and makes the corresponding decisions. A competitive leadership style, in turn, implies a system in which the leader deliberately aligns certain parts of the administration with each other and sets others against each other, simultaneously using a number of alternative and rival information channels in order to obtain the most complete picture of events and policy options. A collegiate leadership style, on the other hand, is based on the idea of different administrative departments representing a relatively open space for debate, in such a way that the different parts of the administration become think tanks. The aim of this style is to achieve the most innovative and best possible solution.

Alternatively, Post (2004; cf. Drennan, McConnell and Stark, 2015, pp. 180–183) has suggested that the different personality types of political leaders are crucial in crisis response. A compulsive personality type is a good organiser and a rational problem-solver, but when faced with a crisis, this type of leader has a tendency to panic because an unexpected event takes them out of their comfort zone. A narcissistic personality type is extremely self-confident and unfazed by making tough decisions when confronted by a crisis. The inherent danger in this type of leader is the tendency to be too quick to judge without considering all the alternatives. A paranoid personality type, on the other hand, is always suspicious of others. In a crisis situation, this paranoia becomes exaggerated, with the tendency to interpret the crisis as a part of a long-standing conspiracy.

5.3 Crisis decision-making theories

Notwithstanding the practical-normative literature, the issue of the nature of crisis decision-making is one of the most theorised subjects in the field of crisis management. The task of this debate is not so much that of proposing normative rules for decision-makers, but rather trying to explain decisions through a variety of theories. Nevertheless, this theoretical debate reveals some interesting points about how crisis decision-making may occur and how it deviates from normal-time decision-making.

Such studies on crisis decision-making do not form a neatly distinct tradition. Although decision-making has been regarded as a central component of studies on crises – some theorists indeed include the necessity to make critical decisions in the very definition of a crisis (Rosenthal, Charles and 't Hart, 1998) – and although studies on decision-making often focus on crises, both issues can still be approached from a number of perspectives. Consequently, there is a rather impressive body of literature on crisis decision-making, especially in the field of international politics, focusing largely on high-politics crisis decision-making in matters of war and peace. Hermann's work (1963) and Allison's study (1971; see also Allison and Zelikow, 1999) on the Cuban missile crisis form nodal points for examining crisis decision-making, while Roberts (1988), Janis (1989), Holsti (1989) and Brecher (1993) all cover this broad area by linking central themes. Literature of this sort often focuses on developing and testing theories on decision-making through single or comparative small-n case studies.

In the European context, a noteworthy movement is the so-called Leiden-Uppsala School of crisis management, which prefers cognitive-institutional models over rational models. This literature focuses explicitly on political leadership as the main crisis management actor, with an emphasis on civilian crises (e.g. 't Hart, Stern and Sundelius, 1997, 1998; Stern, 1999; Stern and

Sundelius, 2002; 't Hart and Sundelius, 2013; Boin, 2004; Boin et al., 2005). There is also a more heterogeneous body of literature in which crisis decision-making is usually understood as only one of the focal areas in the wider crisis management field related to public administration (e.g. Drennan, McConnell and Stark, 2015; McEntire, 2015; Deverell, Hansén and Olsson, 2015; Farazamand, 2014; Fagel, 2013; Haddow, Bullock and Coppola, 2011). In this literature, a great deal of attention is paid to civilian emergencies and disasters, rather than issues of military security, and it is often rather practice-oriented, with links to the operative and planning levels.

Further, we can identify a body of literature which deals with corporate crisis-management issues (e.g. Crandall, Parnell and Spillan, 2014; Dezebhall and Weber, 2011; Regester and Larkin, 2008; Harvard Business School, 2004; Fink, 2002). This genre usually focuses on reputation management, however, and – although closely related to decision-making – it primarily discusses decision-making in terms of crisis communication matters.

All these viewpoints can, however, be discussed in terms of the rival explanatory theories or rather schools of thought on (crisis) decision-making, which is the task of the current section. These are summarised in Table 5.1.

To start with, much of the debate has centred on the theme of rationality in crisis decision-making. Needless to say, rationality is a multifaceted concept. A strict methodological definition of the term – as proposed by positivist science – is that while real life always diverges from a rational idea, a theory should not treat persons as individuals with their own psychology and preferences, but rather seek to turn individual behaviour into that conducted by anybody in a similar situation (Popper, 1994). Yet rationality can also take individual preferences and beliefs into account. The mere notion of instrumental rationality does not predict anything about what an actor should want in a choice situation. We can easily imagine a crisis situation in which two different crisis managers would have divergent goals rooted in different belief systems. In this instrumental sense, the rationality criteria do not include an evaluation of the goals per se. Rationality is merely understood as a process of gathering sufficient information about the costs and benefits of the available action alternatives and then choosing the option most likely to lead to a better outcome, according to the pre-existing desires among the available alternatives (Elster, 1986).

Crisis decision-making may follow this ideal procedural or instrumental model of rationality. This normative rule is often taught to students preparing for the so-called sharp-end professions (e.g. first responders). When they face an accident or an emergency, they should first define the situation, define their goals (which might be obvious, e.g. saving life before property), define the available options towards achieving that goal (e.g. different rescue routes, or the set of equipment to be used), define the probable outcomes of the rival options (e.g. one option may be quicker and more lives could be saved, but it could also be more risky), and then make a choice that best matches their goals. However, while in a normal situation one could weigh the pros and

Table 5.1 Decision-making theories

Main theoretical decision-making schools	Unit of analysis	Main assumptions
Rational choice	Any actor (individual, government, etc.) that can be treated as a unitary actor	Based on one's beliefs and goals, when faced with a decision-making situation, an actor identifies the available options, considers the probable outcomes of the rival options (expected utility value), and then makes a choice that best matches one's goals.
Bounded rationality	Any actor (individual, government, etc.) that can be treated as a unitary actor	The decision-maker is not an 'optimiser' but rather, a 'satisficer': one seeking a satisfactory solution based on imperfect information, and in the presence of cognitive biases. Thus the decision-maker stops considering alternatives when he thinks that he has found a satisfactory solution.
Bureaucratic policy	Small formal groups of decision-makers (e.g. the government)	The preferences of those contributing to the decision compete with each other, and the materialised outcome is the result of a campaign by representatives of various institutions with different interests and different degrees of power, of their compromises, bargaining, conflicts, mutual confusion, bewilderment, and so on.
Organisational routines	Decision-makers within organisations	More or less fixed organisational routines, standard operating procedures and programmes that not only help decision-making at any given time, but may also constrain behaviour and limit the range of options to choose between.
Cognitive-psychological theories	Individual, small groups	An important factor is a person's view of the environment, which varies from person to person according to their belief systems, operational codes, values, state of mind, attitude, emotions, age, physical health, experience, knowledge, stress, and so on. In small groups, such phenomenon as groupthink may lead to a situation where an urge for consensus rather than choosing the best alternative becomes the most important goal.

cons of different options carefully, a crisis situation is characterised by uncertainty and lack of knowledge.

Normally, when given a rational choice situation, we can talk about both parametrical and strategic choices. In the former case, an actor is under external limitations that are mostly parametrical or given, and has to assess these restrictions to the best of his ability before deciding what to do. What aggravates a strategic choice situation, on the other hand, is a mutual dependency of decisions; before making a choice, an actor must assess the possible actions of others, which may call for an estimation of how others assess the actions of the actor himself. A foreign policy crisis is more often a strategic rather than a parametric choice situation, as are most company reputation crises, counter-terrorism operations, financial crises, and so forth. Even in many civil emergencies, a crisis manager has to take into account the reactions of the people to be rescued under different strategic options (e.g. one does not want to create panic and chaos by opting straight for an evacuation). In this strategic interaction, actions are chosen both for their immediate effect and for the effect they have on the other actors' choices. While in a crisis situation, one rarely relies on any rational choice theory – simply trying to reason in a rational fashion instead – some of the related techniques incorporate simulation techniques whose aim is to predict the outcome of, say, a complicated political situation under different action options (e.g. Bueno de Mesquita, 2001). In the best case, these could be used as decision support tools for rational crisis decision-making.

Bounded rationality

A crisis situation seldom allows fully rational decision-making, however. Bounded rationality is already a step towards taking into account psychological factors as it emphasises the improbability of optimal choice in most cases. The decision-maker is, rather, a 'satisficer': one seeking a satisfactory solution based on imperfect information, and in the presence of cognitive biases. Thus the decision-maker stops considering alternatives when he thinks that he has found a satisfactory solution, even if the optimal solution were the next one to be considered (Simon, 1947; Bendor, 2010). This model fits rather well when explaining crisis decision-making, which is characterised by time pressure and often a lack of information, not only about the characteristics of the crisis but also about the existing action alternatives and their potential outcomes.

The solution to these obstacles for optimising rationality is discussed within this school of thought. One way to enhance the quality of crisis decision-making, according to the bounded rationality school, is to expand the decision-making body. One decision-maker inevitably views any crisis situation through a rather narrow lens and is a prisoner of their previous experiences. This might pose an obstacle to finding the best solution as one does not see all the options. The aim is to consider any possible interpretation of the situation and not to ignore any alternative action strategy. A somewhat similar way to improve an individual's unavoidably biased decision-making capacity is to rely on the organisation. A larger organisation, according to the bounded rationality theorists, allows the individual to get closer to objective rationality by bringing on board such advantages as division of labour and resorting to a number of experts when preparing for decision-making, which in turn allows multiple complex processes of simultaneous processing.

Various models of decision-making can also be combined. For example, the so-called poliheuristic decision-making theory (Mintz and Geva, 1997; Mintz, 2004; Dacey and Carlson, 2004) claims that a person will never see all the possible options and their outcomes. Instead, due to the cognitive biases related to the processing of information, decision-makers orientate towards only a limited number of policy or action options. Decision-making takes place in two stages. First, decision-makers close their minds to some of the options, consciously or unconsciously, based on their normative and cognitive preconditions. Only within this limited set of options do they apply the cost–benefit calculations in accordance with the rational choice. This approach is based on the idea of combining the cognitive school of thought with the rationalist school of thought, much like the bounded rationality school does. Furthermore, decision-makers do not take into account the decision outcomes in all their possible forms, for example, by considering their political, economic or military consequences in some mathematical way as the rational choice theory presupposes. In contrast to the rational choice model, which invites the decision-maker to calculate the cost–benefit net asset values of the various options, dissecting all the possible dimensions of an action, the poliheuristic model teaches that in reality the decision-makers are not thinking in terms of compensating for a loss in one dimension (e.g. economic) with a benefit from another dimension (e.g. political). Instead, they set a kind of bar for all dimensions above or below which they will not go, even if the net benefits of such a decision exceed the disadvantages (Mintz, 1993). In short, the poliheuristic theory expects people to act rationally, but based on their experiences they apply heuristic short-cuts or some kind of rules of thumb.

Bureaucratic roles and organisational routines

Another model or theory challenging the rationality of decision-making is the so-called bureaucratic politics model, popularised in particular by the above-mentioned well-known study by Allison on US decision-making during the Cuban missile crisis in 1962 (Allison, 1971; Allison and Zelikow, 1999). It stems from the premise that even in a system of decision-making dominated by one person, such as the president, the decisions are not taken by this one person alone, but collectively, surrounded by other high-level actors, aides and consultants, and the final decision-maker is to a considerable extent dependent on the behaviour of other bodies in the administration. The objectives and interests of a bureaucratic actor are to a considerable degree related to their bureaucratic position. This is due to the fact that members of a bureaucracy are inclined to believe that it is the well-being of their bureaucracy that is particularly crucial from the general perspective. In turn, the well-being of a bureaucracy is to a large extent dependent on the preservation or enhancement of its own influence, on the fulfilment of its remit, and on securing the necessary resources. The bureaucratic position simultaneously facilitates and places limitations on behaviour, providing certain resources, but also creating normative roles corresponding to various positions.

Hence, the bureaucratic politics model questions the rationality of decision-making. It is not that there are many people taking part in the decision-making process (which is almost self-evident), and not even the fact that these people represent some kind of bureaucracy or promote their own interests. It is more about the fact that a materialised outcome is not the result of a single choice made by a rational individual or a coherent group, but is always some kind of a blend of several actors' choices. The outcome is not rational in any logical sense, even though there may happen to be rationally behaving individuals among the decision-makers. In the bureaucratic politics model, the preferences of those contributing to the decision compete with each other, and the materialised outcome is the result of a campaign by representatives of various institutions with different interests and different degrees of power, of their compromises, bargaining, conflicts, mutual confusion, bewilderment, and so on.

Constructive criticism (e.g. Smith, 1980; Bendor and Hammond, 1992; Welch, 1992; Alden and Aran, 2012, pp. 34–43) of the bureaucratic politics model has given rise to the notion that this model should not be interpreted in such a way that the mere position of the actors dictates behaviour. The precise constituents of each role and the relations between them differ from case to case. Moreover, every individual has several roles at any given moment in time. Individual actors always represent several groups: a certain bureaucracy, an economic or other interest group, a party, and so on. On the other hand, a bureaucracy or a group can comprise far smaller groups that seek particular benefits. Roles linked to positions are also open to criticism, punishments and changes. Seldom are roles so strict and specific as to leave no room for interpretations and variations at the personal level. The problem with this type of flexible interpretation of the bureaucratic politics model, however, is that it makes the use of the concept of bureaucratic interest problematic; one simply ends up with a situation where the boundaries of different interests are so blurred and obscure that it is impossible to draw conclusions on the basis of a mere bureaucratic position.

In the aforementioned book by Allison (1971), another model was also presented, challenging the rationality of decision-making. In its original form, the organisational process model described politics not as deliberate choices but rather as outputs of large existing organisations functioning according to standard patterns of behaviour. While rational choice theory and the bureaucratic politics model concentrate on the very unit or group of decision-makers, usually on the choices of the leaders, the emphasis of the organisational process model is on more or less fixed organisational routines, standard operating procedures and programmes that not only help decision-making at any given time, as the bounded rationality school argues, but may also constrain governmental behaviour and limit the range of options to choose between. While the model outlined by Allison clearly focuses more on pre-decision constraints arising from the complex organisational character of state machinery, it can be extended to post-decision developments as well (cf. Bendor and Hammond, 1992). In larger institutions, almost any decision must be implemented by lower-level organisations that follow routines and procedures that may cause unintended consequences. On the other hand, it has been seen that not only routines but also the breakdown thereof can sometimes pose the greatest threat to successful management and governance, particularly in crisis situations.

In any event, organisations are based on norms, a repertoire of institutionalised practices, and possibly plans as well as resources that are at the disposal of decision-makers, but which may also serve to constrain them. The organisational or institutional framework also becomes important in terms of the very decision-making situation, not least because the actors that make crisis decisions are usually not only constrained by organisational procedures, but also selected in part because of previous institutional settlements. It is actually here where the bureaucratic politics model and the organisational process model meet. Decision-makers do not appear on the scene out of the blue. The casting is embedded in the institutional complex. Organisations entail positions and mandates that tend to determine the group of decision-makers that are activated, particularly during unexpected situations such as crises.

Cognitive and psychological biases

Cognitive and psychological theories are traditionally seen to rival the idea of rationality most strongly. A key challenger to rational choice theory within these theories is prospect theory (Kahneman and Tversky, 1979; Levy, 1997; McDermott, 1998). The theory, proved by means of laboratory tests within a number of contexts and cultures, argues that people are generally risk-averse in choices involving gains and more likely to be risk-seekers in choices involving losses. As many choice situations can be framed in terms of both gains and losses, this notion challenges the fundamentals of rationality; instead of choosing on the basis of the outcome's utility value, the way the choice's starting point is framed affects the preference order.

Decision-making in small groups, as is often the case in a crisis situation, places additional limits on rationality. Small groups tend to reduce uncertainty and produce more straightforward decisions than individuals would have done separately. This so-called groupthink (Janis, 1982) may also lead to an overemphasis on common views, with the result that an urge for consensus rather than choosing the best alternative becomes the most important goal. In these circumstances, a small group of decision-makers, no matter how astute, can reach not only sub-optimal but clearly irrational decisions on the most important issues. If the small group works under groupthink, they may even omit clear facts and not allow any contradictory information and opinions to interfere with the group dynamics.

The opposite of groupthink is so-called polythink, introduced by Mintz and DeRouen Jr. (2010, pp. 49–54) when studying foreign policy crisis decision-making. Polythink occurs when a decision-making group is extremely non-uniform, in which case its views, preliminary expectations, comments and reviews of the decision-making options and criteria at hand differ significantly from each other. A diverse decision-making group naturally precludes groupthink and tunnel vision. However, when decision-making is based on the lowest common denominator, it leads to conflicting interpretations about what is at stake and what the outcome might be. Information leaks outside the group are also probable. Such a situation may lead to a decision-making group sending inconsistent signals to their counterparties or their own supporters.

As mentioned above, rationality is usually understood to include the notion that people gather sufficient information to form their desires, beliefs and assumptions, and then act accordingly when choosing between the available alternatives (Elster, 1986). However, psychological and cognitive tests, when applied to (crisis) decision-making studies in particular, have proved that a person unavoidably overlooks some of the alternatives for several reasons, or refuses to recognise given information (Stein and Welch, 1997; Vogler, 1989). At the heart of the problem is a person's view of the environment, which varies from person to person according to their belief systems, values, state of mind, experience, knowledge, and so forth. The facts pertaining to various situations are not perceived in the same way by everybody; the facts do not 'say' or 'tell' anything in and of themselves. Instead, a person has to select them, put them in order and classify them to give them meaning, and finally to act on the basis of the knowledge duly accumulated.

In other words, information processing is selective. Striving for consistency in the adopted attitudes or belief systems can lead to situations whereby a person closes their mind to controversial information. Faced with critical information, a person is inclined to misunderstand it, deny it or leave it wholly unprocessed. If one's trust in a presupposition is strong, it is also possible that a person will be inclined to make a decision before they have sufficient information at their disposal. All information is interpreted whenever possible in such a way that it supports existing beliefs, although it could just as easily support any other interpretation of the situation. When this is eventually recognised, a person often chooses the first alternative that offers a way out, and instead of radically changing their views, they merely add some exceptions and superficial changes to their existing outlook.

Information processing becomes particularly challenging in situations that cause stress, such as a crisis, being faced with an unmanageable workload, anxiety or fatigue, and so forth. For instance, in many early warning cases, the issue is one of deciding whether or not to act on the basis of available information under conditions of uncertainty. The decision to act may incur significant costs, such as shutting down a nuclear reactor or mobilising a number of rescue forces, but so might inactivity. Early warning signals and the information a crisis manager receives when confronted with a crisis situation are seldom unambiguous and there are often incentives to interpret the available information in various ways, while making cost-benefit calculations in the midst of a stressful situation. Cost-benefit calculations do, however, demand the rational weighing of the pros and cons of the alternative action lines. Yet stressful situations are ones in which a person is particularly prone to behaving irrationally. It has been argued that in such situations the number of misjudgements and miscalculations rapidly increases, risk-taking inclinations become more frequent, the consideration of relevant facts becomes highly selective, the difference between what is essential and what is irrelevant disappears, the ability to abstract meaning from details weakens, and it becomes more difficult to tolerate complexity (Flin et al., 1997; Nicholson, 1992, p. 127ff.). This adds plausibility to the argument that while rationality may be at work in normal circumstances as some kind of basic point of measurement, it does not work in crisis situations because people behave differently in a crisis and may often be unaware of the logic of their choices.

On the other hand, Forsberg and Pursiainen (2017) have argued, referring to Russia's decision to invade and then annex Crimea in 2014, that while some external (power politics) and domestic factors (fear of colour revolution) may provide the necessary rational conditions for this decision, they are not sufficient. Psychological factors have to be taken into account in order to understand the decision. Thus, psychological factors would work in a way not as total explanation but as triggers or intervening variables that explain the outcome. Indeed, even the advocates of the rational choice methods (Morrow 1997, pp. 29, 30) admit that psychological approaches perform "at least as well as rational choice methods, because the former accept all the variables that the latter recognize as important in explaining a case." Rationalists, however, do not consider this as a sign of theoretical superiority, rather only as a proof of greater "descriptive accuracy of an individual case".

5.4 Crisis communication

Crisis communication is a multifaceted concept. It cannot be limited to the response phase of crisis management. Crisis communication is a necessary horizontal factor in all crisis management phases, yet it is most natural to discuss in the context of response. As Hale, Dulek and Hale (2005) concluded after reviewing the literature, the early work on the response stage of crisis communication focused on channels and information flow, dealing with these issues mainly in terms of crisis decision-making. This research concluded that while a crisis leads to an excessive need for information, the number of communication channels that are used tends to decrease, leading to information overload and channel bottlenecks. Yet, the crisis communication literature soon started to emphasise the external dimension of crisis communication, mainly concerning the crisis manager's communication with the media, the public and other stakeholders.

In order to cover most of the crisis communication elements, in this section we will discuss four issue areas: (early) warning from the crisis communication perspective; the coordination and communication of actions in cases where there is a network of interested parties rather than just one organisation as a crisis manager; external crisis communication in the above-mentioned meaning; and internal crisis communication within an organisation.

Early warning as crisis communication

Crisis communication includes the reception and interpretation of a crisis signal. In the previous chapter, early warning was already dealt with from the preparedness perspective, emphasising that one should understand monitoring, early warning and a warning or an alert as a complex early warning system (EWS) rather than as a one-time activity. If that system works, some crises could be prevented altogether. In any case, EWS concerns the temporal dimension of warning about a potential or an impending problem, which, if successful, should help to prevent or mitigate harmful consequences, or provide leeway for at least some preparatory or mitigating measures, such as evacuation or preparing the necessary equipment for an early response.

There is no factual difference between the concepts of early warning or mere warning in the context of an actual emergency or a crisis, in the sense that an early warning signal can precede, or sometimes immediately follow, the actual event, thus helping to prevent or mitigate the emergency. In understanding early warning as an integrated system in crisis management rather than as a one-time activity, the early warning time span becomes considerably longer and the whole concept much more multi-dimensional than if we were to understand it strictly as acute crisis communication. Early warning, for instance, with respect to climate change consequences, can occur over a period of decades. Time series data collected over several decades have provided an impetus for preventive and mitigating actions, such as changes in human behaviour, energy policy or infrastructure design. This can be understood as a preventive and response strategy to a typical creeping crisis. On the other hand, a few seconds or the last few minutes before an imminent danger or disaster can make a huge difference in some cases. In other cases, the first few minutes or hours after a disaster might be the most important in terms of reducing its impact on people, the environment and property. Early warning might be needed even after the initial disaster phase. In so-called complex disasters, there is a high probability that a natural disaster will be followed by a technological, health or socio-economic disaster, or that an environmental or economic catastrophe could ensue after a technological disaster, for instance. This probability, properly understood, may provide room for preventive action or help in preparing for potential successive disasters.

However, early warning is possible, and early warning strategies make sense only if there is an early signal that can be detected, in order to prevent, mitigate or be prepared for a predicted problem. The problem is that "signals and messages associated with threats are often faint, subtle or not easily detected, and, in addition, are often incorrectly interpreted" (Sellnow, Seeger and Sellnow, 2013, p. 51). In the conflict early warning literature, it has been suggested (e.g. Vetschera, 2005), however, that every future societal crisis will generate signals at an early stage and should not surprise a close observer with the necessary detailed knowledge. Even when unexpected discontinuities do occur in the context of social systems, they do not happen overnight, but invariably announce their arrival in advance. The issue is to recognise faint or weak signals, so as to be able to transform mere factual information into an early warning signal of a forthcoming crisis. The closer a signal is to the outbreak of a crisis, the more specific the signal will be and thus the more easily identifiable. As the critical issue often hinges on having sufficient time to react, early warning bottlenecks primarily emerge as factors that shorten reaction times.

Chapter 2 discussed how systems often have nonlinear dynamics whereby the behaviour of the system is not causally deterministic, but rather random, and where the randomness is caused by the exponential growth of errors and uncertainties (Bellavita, 2006; Gundel, 2005; Klinke and Renn, 2002). In these cases, detecting a signal of a future crisis is virtually impossible. Some crises in turn happen so quickly and unexpectedly that there is no room for early warning. On the other hand, crises are often extremely difficult to identify even if it were possible. The question that then arises is how to detect the signal amid the surrounding noise. This is particularly true if it is a question of the above-mentioned creeping crises or completely different types of crises that have occurred previously. As for creeping crises, Lagadec (1997, p. 25) has summarised the phenomenon as follows:

> The crisis may inch forward and take over the entire field of operations, through 'underground' channels that the organization fails to recognize. What is missing is the characteristic feature of an emergency: a clear trace that would justify triggering the warning procedures and mobilizing resources. The crisis only makes itself known once it has established solid bridgeheads. Then, it is often too late to counter-attack.

There are several factors that are crucial for the success or failure of early warning (Pursiainen, 2007). Human factors limit a person when it comes to receiving and processing information that contradicts existing beliefs and values. These limitations can be dealt with by using systematised 'imagine-the-unimaginable' scenario-building during the risk-assessment phase, as well as by enlarging the group responsible for risk assessment and monitoring crisis signals. Similarly, historical experiences may constrain or narrow the identification of new threats or result in the wrong cognitive short-cuts, leading to an inaccurate diagnosis, and consequently to an inappropriate or delayed response. These could be avoided to some extent by awareness-raising and training the relevant actors to take this particular problem into account. The more ergonomic the situation (un)awareness problems, already briefly discussed in Chapter 3, the more they relate to a human being's physical and psychological limitations in interacting with technology, and thus in identifying the early warning signals. These can be diminished by prolonged human-factor testing and, naturally, by the effective selection and training of high-tech system operators.

The role of technology in communication is crucial. Remote sensing systems, satellites, geospatial technologies (e.g. Ferrara, 2009; Gao et al., 2009), and so forth, can confer huge benefits on early warning and, in an ideal case, one should harness all the available technological know-how in order to obtain information on risks. This means that technological detection systems should be in place; the connections between the detectors and the people/property in danger should be working; and, when available, one should make use of automatic response systems. Yet while exploiting technological and high-tech solutions in early warning is recommendable, the flipside is that too strong a dependence on technology should be avoided so as not to render the system vulnerable to interdependent technology failures.

Early warning failures often result from a poor safety culture, discussed in some detail in Chapter 3. A checklist to keep in mind when evaluating a system is that a good safety culture means giving the highest priority to safety over other factors; the organisation should be willing to raise safety concerns and to learn from errors, incidents and accidents; the safety values, attitudes, perceptions, competencies and patterns of behaviour should be shared among all members of an organisation, also considering safety concerns from 'below'; and recognising that a good safety culture also entails the willingness to submit to outside inspection and to share safety-related information with the public.

Many early warning failures can be traced back to organisational and legislative factors. In the public policy sector as well as in high-risk industries, the basic rule is that an early warning system has to be institutionalised and regulated by legislation. It should be characterised by a clearly defined mandate, tasks and goals, clear divisions of responsibility, clear leadership and decision-making structures, clarity in the chain of command, clear communication systems within and between relevant organisations, institutionalised post-disaster evaluation and learning systems, as well as the maintenance of training and financing, and the upgrading of equipment. The prevailing early warning approach to disaster reduction states that while well-regulated international, national and regional emergency management systems should be in place, early warning should be based on community participation and cooperation with the local decision-making, civil-society and private actors. In the public policy sector, political commitment, especially towards investing sufficient economic resources for early warning purposes, is crucial for a well-functioning early warning system. In order to integrate early warning into national economic planning, emergency authorities and specialists should highlight the economic benefits of early warning, targeting senior government and political leaders as well as local decision-makers by using practical cost–benefit demonstration methods, examples and case studies.

Network communication and coordination

Some crises are not managed from within just one organisation but by many centres of decision-makers, which may be horizontally or vertically organised. In larger organisations, there are also numerous departments that mainly focus on their own activities in normal situations, but which have to communicate with each other much more closely in a crisis situation. In the public sector, this is often the case in civic emergencies and disasters. In the previous chapter, we have already touched on the so-called network management model, aimed at the horizontal coordination of decision-making and actions between several agencies, or presupposing a rather open, horizontally organised decision-making model. Boin and Bynander (2014) see coordination not predominantly from the perspective of communication, but rather either as collaboration or a form of directive action.

However, this collaboration needs communication in order to be successful. Indeed, the focus has been on adaptive systems for some time now, which in turn depend upon an information infrastructure that is sufficiently flexible to adapt to the changing conditions. Following Kapucu (2009), the dynamics of crises require a rapid search for, exchange and absorption of information across many agencies, starting from the pre-crisis planning system and taking into account the logic of crisis management across multiple agencies and jurisdictions. This notion is in line with criticism of the linear, hierarchical command-and-control system, but emphasises the need for a more network- and partner-based system, including public–private and non-profit partnerships.

Boin and Bynander (2014) have also noted that the larger and more complex the crisis, the more likely it is that a planned command-and-control response will not work. Instead, a response network, consisting of many organisations each with their own responsibilities and ways of working, will emerge either by default but usually by improvisation. Some governance models propose coordinating multiple decentralised response organisations under a temporary hierarchical structure (Moynihan, 2009), which seemingly works well and is supported by practitioners in such countries as Sweden, for instance (Wimelius and Engberg, 2015). Drennan, McConnell and Stark (2015, pp. 174, 175) and Parker and Stern (2002) believe that coordinating this kind of network crisis management is usually achievable, but it demands some kind of agreed coordination and negotiation body beforehand.

Hence, these biases of the command-and-control system are often resolved with a coordinating actor, or a coordinating body that is responsible for having a holistic, cross-sectoral overview of the cross-organisational communication routines, and that will correct them if necessary. If the early warning system is not integrated and coordinated, the early warning signal may not be received or identified even if the monitoring system as such is in place. For instance, two organisations may have individually important information, but it will only become an early warning signal if these two information sources are combined. Therefore, it is generally agreed that an integrated surveillance and information system should exist to ensure that the collected data is properly shared, analysed and processed to enable the recognition of possible early warning signals. When the crisis has already been recognised during the decision-making process, it is similarly important for all the relevant information to be made available in order to better understand the alternative options and their probable consequences.

In other words, organisations are often too closed with no mechanism for automatic information flow between authorities. This may be the result of many things such as a basic organisational design flaw, overlapping responsibilities with no clear idea of decision-making chains, organisational and bureaucratic rivalries, and biased practices and routines. As exemplified in the failure to prevent the September 11, 2001 terrorist attack in the US, the diverse and sometimes even rival organisational goals, approaches, culture and structure of different agencies may account for the reluctance of agencies to share information with each other (Parker and Stern, 2002, p. 611).

With a focus on civil emergencies, Bharosa, Lee and Janssen (2010) have noted that communication and coordination problems have different levels. At the community level, typical problems include organisational silos, no incentives for horizontal information-sharing, conflicting role structures, a mismatch between goals, independent projects, a lack of standardisation or interoperability, and heterogeneous systems in general. At the agency level, the very issues that may also produce positive results, such as reliance on protocols, communication between predefined contact persons only and focusing on vertical information-sharing, can become problematic for optimal information-sharing in a multi-actor environment. Communication problems are also obvious at the individual level. Such issues – already discussed in part in other contexts above – that pose a challenge to communication include information overload, the inability to determine what should be shared, the misinterpretation of information, the prioritisation of one's own problems and access limitations.

External communication

External communication typically refers to communicating with the media, external stakeholders and the general public. The starting point is understanding the importance of this communication. As discussed above, the media are often not only the first to pick up crisis signals, but may also foster the crisis feeling, as media representatives often define the values at stake, and create time pressure and a sense of uncertainty (Forsberg and Pursiainen, 2006). Moreover, badly managed external communication not only adds to the crisis feeling, but may also create an additional crisis – a media crisis – which can turn out to be the main issue in crisis management.

A public organisation, such as a rescue organisation or authority, usually has an external communication task in a crisis, whose goal is to defend stakeholders or the public at large against the consequences of the crisis. From the perspective of another type of organisation, such as a profit-seeking company, communication serves to reduce the harm caused by these crisis episodes to the organisation itself. It is a question of maintaining the organisation's reputation during a crisis and preventing reputational damage after a crisis event. This reduces the crisis response mainly to external communication, that is, "crisis responses as messages designed by the organization with a strategic goal in mind ... to change, alter, or shape perceptions of crisis attributes that influence how stakeholders view the organization in a crisis event" (Fediuk, Pace and Botero, 2012, p. 222).

Thus, the external communication serves different or several purposes depending on the organisation, and naturally also on the nature of the crisis. While these roles and purposes may be mixed and overlapping, no general strategy for external communication applies to all, and has to be tailored instead. It has also been argued that in a reputational or image crisis, a history of similar crises intensifies the reputational threat of the current crisis (Coombs, 2004), and hence should be taken into account in the communication strategy. In any case, some alternative strategies have been proposed, actually harking back to the issue of framing discussed above. Coombs and Holladay (2006), for instance, suggest that there are three communication options. First, one can deny the existence of the crisis. Second, one can alter the attributions of the event to make it appear less negative to stakeholders. And third, one can alter how the organisation is viewed by stakeholders.

In much the same vein when discussing crises with political components, Boin, 't Hart and McConnell (2009) have distinguished between three basic options. First, denial that the event in question represents more than an unfortunate incident. Second, deeming the event to be a critical threat to the collective good embodied in the status quo that existed before the event came to light. And third, deeming the event to be a critical opportunity to expose deficiencies in the status quo ex ante. Benoit (1997), for his part, has concentrated on the image repair strategies of a business organisation, and has identified further possible communication strategies with real-life examples, the main categories being denial, evasion of responsibility, reducing offensiveness, corrective action and mortification (apologising).

Claeys, Cauberghe and Vyncke (2010), in turn, have tested Coombs and Holladay's (2002, 1996) argument that crisis response or communication success depends on the type of crisis in question. They conclude, in effect, that matching crisis types and crisis responses does not lead to a more positive perception of a company's reputation than mismatches. Further, they differentiate between three types of crises: victim crisis, accidental crisis and preventable crisis. They then define three types of crisis response or communication strategies: the deny strategy, the diminish strategy and the rebuild strategy. They have found that preventable crises have the most negative effects on organisational reputation and that the rebuild strategy leads to the most positive reputational restoration.

There are, however, several best practices and normative guidelines for different fields, such as public administration (e.g. McEntire, 2015, pp. 204–213; UK Cabinet Office, 2013; Haddow, Bullock and Coppola, 2011, pp. 139–174), and businesses or organisations (e.g. Regester and Larkin, 2008, pp. 182–198; Seeger, 2006; Harvard Business School, 2004, pp. 71–74; Fink, 2002, pp. 92–120), when it comes to dealing with this task in practice. The normative rule is that public and media communications should become an integral part of the strategic crisis decision-making arrangements.

From a purely practical point of view, one of the first tasks in any crisis is to establish a Crisis Communication Team, which should be staffed with persons from the respective organisation(s) at a high enough level. Its task would be to agree on a communication plan that ensures consistency in the information provided by the organisation and, if there are several organisations or state agencies involved, to coordinate the message between them. This cross-agency coordination of information should be established at an early stage after an incident. Similarly, one of the first issues to tackle is sometimes called positioning. The rule of the thumb is: Tell it all, tell it fast and tell the truth! Experience has shown that this maxim, namely that one should never try to lie, deny, hide or ignore the situation, is the winning strategy. The external crisis communication should be proactive, even when one does not have a clear picture of the characteristics of the crisis. The first news release should include at least who, what, when and where, and then add that the issue is under investigation and more information will be forthcoming. In addition to press releases in print and electronic form, regular briefings from the senior personnel involved are required to quash rumours and conjecture. One should refrain from speculating, especially about causal factors, and one should not present half-promises that raise expectations. This is not always easy. As Sellnow, Seeger and Sellnow (2013, p. 30) have remarked, "the media often seek immediate explanations regarding cause and blame and disseminate this information very broadly". Therefore, the crisis manager's remit includes keeping a close eye on media coverage, and correcting inaccurate reporting immediately.

When it comes to the executive work of the communication team, one should designate a spokesperson or press officer, who would have the primary responsibility for making official statements and answering media questions throughout the crisis. Such persons often have a journalistic background. To be successful, the person should be experienced and skilled in handling the media, able to establish credibility, remain calm in stressful situations and demonstrate awareness of the needs of the media, including the need to meet copy deadlines or broadcast live reports. In the event of major crises, this person needs a backup team of press officers and coordinators, technical experts, advisors and liaison persons to organise media visits to any scene, including transport arrangements where events have occurred in a remote location. The spokesperson should be able to control the media to some extent, on the basis of trust, for example, by agreeing to disclose or withhold certain details for the time being (e.g. during hijack situations or hostage negotiations).

In the event of civil emergencies, tasks may include organising a venue where the media can be based, including their equipment. This should be close but not too close to the crisis location to ensure that the media is not in the middle of the action, or in the expectation that they will simply move to a more accessible location independently. One should be consistent when it comes to sticking to rules established with the media during the crisis, and by treating all media equally. While aiming at controlling any possible negative impact of the media on crisis management as far as possible, one should not give the impression that the media are not being permitted to do their job.

Press conferences at the beginning of and during a crisis, often broadcast live, are particularly important in ensuring that the right message is being relayed to the public. The spokesperson should master the situation, setting time limits and handling questions prudently. The responsible persons may start with a short, prepared statement. If they are inexperienced, one should use some time to predict and practise dealing with tough questions before the press conference begins. General communication skills should be appreciated and followed such as speaking loudly and clearly, establishing eye contact with the audience, remaining friendly, cool-headed and confident, answering only the questions that are asked, talking about people first, property second, and money third. One should focus on one's feelings about the situation, and how it will be prevented from happening again. Should one not be able to answer tough questions, one can simply say 'I don't know the answer to that yet' or 'Thank you for the question. I will have to check that'. A typical mistake is assuming that journalists cease to be reporters when the press conference is over. Therefore, one should never speak off the record. Similarly, one should never try to go over a reporter's head to stop a story, as this will probably only serve to exacerbate the situation.

The media is often the most important channel for communicating with the public at large. However, in today's world, much direct communication takes place through the internet. Therefore, it is of the utmost importance to update website information regularly. This necessitates a web team that can monitor news sources and feeds for breaking news and additional information. Experience dictates that in a crisis situation such as a nuclear or radiological emergency, the relevant websites tend to crash just when they are needed the most. Therefore, it is important to ensure in advance that one's website has the capacity in place, or can quickly gain the capacity, to deal with the huge number of hits that a crisis or emergency might generate. In larger crises, one should have a designated team to deal with the public directly, including a social media team. In civil emergencies, citizens often contact an authority or an organisation directly by phone or email, for instance, to receive information about when the electricity supply might be restored, or to receive news about relatives who might be caught up in the disaster. Responding to inquiries promptly – even by saying that information is still being gathered and that the person will be informed later – is one of the duties of crisis managers in these types of crises, and the failure to do so may come at the expense of criticism and a loss of reputation.

Interactive social media sites and other new communication-related technology have both offered new tools for crisis management in terms of public communication – and communication from the public to crisis managers – but have also given rise to a new set of problems. Studying the use of Twitter in the context of the Great Tohoku earthquake followed by the devastating tsunami in Japan in 2011, Acar and Muraki (2011) found that people in the directly affected areas tended to tweet about their unsafe and uncertain situation while people in remote areas posted messages to let their followers know that they were safe. That particular analysis revealed that unreliable retweets on Twitter were the biggest problem that users faced during the disaster. Similarly, Dugdale, Van de Walle and Koeppinghoff (2012) found that the information obtained via social media or through text messages during management of the Haiti 2010 earthquake crisis proved particularly useful at the aggregate level, but led to problems such as information overload and processing difficulties, variable speeds of information delivery, managing volunteer communities and a high risk of receiving inaccurate or incorrect information. A survey on the so-called crowdsourcing, citizen sensing and sensor web technologies (Boulos et al., 2011) shows how useful these new information-gathering tools are in many circumstances, but also highlights the challenges related to information overload, unnecessary 'noise', misinformation, bias and trust issues.

Internal communication

Interest has recently emerged to study internal crisis communication (e.g. Frandsen and Johansen, 2011; Johansen, Aggerholm and Frandsen, 2012). In general, internal communication is about communicating between different levels or units within the organisation, usually between management and employees. In organisation studies, it is widely held that communication within organisations is linked to higher levels of performance and service, generating social capital for the organisation if carried out in such a way that it takes into account the employees' needs (e.g. Ruck and Welch, 2012). In a crisis situation, the need for effective internal communication is even greater, as such a situation can prompt previously satisfied employees to quickly change their opinion about and commitment to an organisation.

Preparedness plans were discussed in the previous chapter, where it was mentioned that the larger the organisation, the more likely it is to have a plan for crisis situations. Whether such plans include an internal communication dimension, however, may (according to a large-scale survey focusing on Danish private organisations and municipalities only) depend on whether the organisation in question is a private or a public one. It was found that the internal elements of the crisis management or preparedness plan were in general less prevalent among public organisations compared to private ones (Johansen, Aggerholm and Frandsen, 2012, p. 277).

Frandsen and Johansen (2011) have noted that practically oriented literature on internal crisis communication is very sender-centric. The focus is on the normative issue of how managers must communicate with employees in a crisis situation, basically in the same spirit as normative guidelines on external crisis communication provide best practices for crisis managers on how to deal with the media and the public. While in the latter case this might be justified in the sense that the 'media and crises' theme – including such issues as how the media creates, frames and interferes in crises – is perhaps a separate theme from the very crisis management point of view, when it comes to internal crisis communication, the link to crisis management is closer.

Strandberg and Vigsø (2016) have pointed out several important issues in their case study that should be taken into account when communicating internally within an organisation during a crisis. The case concerned a municipality where a former employee was accused of large-scale embezzlement. They concluded that while the crisis communication was successful when it came to informing external stakeholders and the media, the internal communication between the management and the employees failed. The employees felt that the management had not presented all the information that they needed, which led to assumptions and rumours. The management delivered the same information to the employees that they had given to the media; yet while the media's need for information was fulfilled, the employees experienced a lack of information. When the management failed to deliver sufficient information to the employees, they created their own truth, based upon their interpretation of the situation, which differed from that of the management. The management also put all the blame on the former employee, whereas the employees considered that the crisis was also due to an organisational culture problem; within the organisation, a lack of commitment and trust served to facilitate the fraud and opened the door to misconduct. When control was lacking, nothing stopped the former employee from committing fraud. The employees duly considered that there was a degree of shared responsibility, and the management was seen as dodging the cultural problems. The general lessons learnt were threefold. First, differences between external and internal crisis communication need to be taken into account. Second, a crisis can reinforce existing patterns within a dysfunctional culture. Third, even in an ostensibly clear case of this sort, one should not single out employees as scapegoats and put all the blame on them.

References

Acar, A. and Muraki, Y. (2011) Twitter for Crisis Communication: Lessons Learned from Japan's Tsunami Disaster. *International Journal of Web Based Communities*, 7(3), pp. 392–402.

Alden, C. and Aran, A. (2012) *Foreign Policy Analysis: New Approaches.* London: Routledge.

Allison, G.T. (1971) *Essence of Decision: Explaining the Cuban Missile Crisis.* Boston, MA: Little, Brown.

Allison, G. and Zelikow, P. (1999) *Essence of Decision: Explaining the Cuban Missile Crisis.* New York: Longman.

Ansell, C., Boin, A. and Keller, A. (2010) Managing Transboundary Crises: Identifying the Building Blocks of an Effective Response System. *Journal of Contingencies and Crisis Management*, 18(4), pp. 195–207.

Bellavita, C. (2006) Changing Homeland Security: Shape Patterns, Not Programs. *Homeland Security Affairs*, 2(3). Available at: http://www.hsaj.org

Bendor, J. (2010) *Bounded Rationality and Politics.* Berkeley: University of California Press.

Bendor, J. and Hammond, T.H. (1992) Rethinking Allison's Models. *The American Political Science Review*, 86(2), pp. 301–322.

Benoit, W.L. (1997) Image Repair Discourse and Crisis Communication. *Public Relations Review*, 23(2), pp. 177–186.

Bharosa, N., Lee, J. and Janssen, M. (2010) Challenges and Obstacles in Sharing and Coordinating Information During Multi-Agency Disaster Response: Propositions from Field Exercises. *Information Systems Frontiers*, 12, pp. 49–65.

Boin, A. (2004) Lessons from Crisis Research. *International Studies Review*, 6, pp. 165–174.

Boin, A. and Bynander, F. (2014) Explaining Success and Failure in Crisis Coordination. *Geografiska Annaler: Series A, Physical Geography*, 97(1), pp. 123–135.

Boin, A., 't Hart, P. and McConnell, A. (2009) Crisis Exploitation: Political and Policy Impacts of Framing Contests. *Journal of European Public Policy*, 16(1), pp. 81–106.

Boin, A. et al. (2005) *The Politics of Crisis Management.* Cambridge: Cambridge University Press.

Boulos, K. et al. (2011) Crowdsourcing, Citizen Sensing and Sensor Web Technologies for Public and Environmental Health Surveillance and Crisis Management: Trends, OGC Standards and Application Examples. *International Journal of Health Geographics*, 10(67). Available at: http://www.ij-healthgeographics.com/content/10/1/67

Brändström, A., Bynander, F. and 't Hart, P. (2004) Governing by Looking Back: Historical Analogies and Crisis Management. *Public Administration*, 82(1), pp. 191–210.

Brecher, M. (1993) *Crises in World Politics: Theory and Reality.* Exeter: Pergamon Press.

Breuning, M. (2007) *Foreign Policy Analysis: A Comparative Introduction.* New York: Palgrave Macmillan, pp. 87–92.

Bueno de Mesquita, B. (2001) *Predicting Politics.* Columbus: The Ohio State University Press.

Claeys, A-S., Cauberghe, V. and Vyncke, P. (2010) Restoring Reputations in Times of Crisis: An Experimental Study of the Situational Crisis Communication Theory and the Moderating Effects of Locus of Control. *Public Relations Review*, 36, pp. 256–262.

Coombs, W.T. (2004) Impact of Past Crises on Current Crisis Communication: Insights from Situational Crisis Communication Theory. *Journal of Business Communication*, 41(3), pp. 265–289.

Coombs, W.T. and Holladay, S.J. (1996) Communication and Attributions in a Crisis: An Experimental Study in Crisis Communication. *Journal of Public Relations Research*, 8(4), pp. 279–295.

Coombs, W.T. and Holladay, S.J. (2002) Helping Crisis Managers Protect Reputational Assets: Initial Tests of the Situational Crisis Communication Theory. *Management Communication Quarterly*, 16(2), pp. 165–186.

Coombs, W.T. and Holladay, S.J. (2006) Unpacking the Halo Effect: Reputation and Crisis Management. *Journal of Communication Management*, 10(2), pp. 123–137.

Crandall, W.R., Parnell, J.A. and Spillan, J.E. (2014) *Crisis Management: Leading in the New Strategy Landscape*. Second edn. Los Angeles: Sage Publications.

Dacey, R. and Carlson, L. (2004) Traditional Decision Analysis and the Poliheuristic Theory of Foreign Policy Decision Making. *Journal of Conflict Resolution*, 48, pp. 38–55.

Deverell, E., Hansén, D. and Olsson, E.-K. (eds) (2015) *Perspektiv på krishantering*. Lund: Studentlitteratur.

Dezenhall, E. and Weber, J. (2011) *Damage Control: The Essential Lessons of Crisis Management*. Westport, CT: Prospecta Press.

Drennan, L., McConnell, A. and Stark, A. (2015) *Risk and Crisis Management in the Public Sector*. Second edn. New York: Routledge.

Dugdale, J., Van de Walle, B. and Koeppinghoff, C. (2012) Social Media and SMS in the Haiti Earthquake. Paper presented at WWW 2012 – SWDM'12 Workshop, April 16–20, 2012, Lyon, France.

Elster, J. (1986) *Rational Choice*. Oxford: Basil Blackwell.

Fagel, M.J. (2013) *Crisis Management and Emergency Planning: Preparing for Today's Challenges*. Boca Raton, FL: CRC Press.

Farazamand, A. (2014) *Crisis and Emergency Management: Theory and Practice*. Second edn. Boca Raton, FL: CRC Press.

Fediuk, T.A., Pace, K.M. and Botero, I.C. (2012) Crisis Response Effectiveness: Methodological Considerations for Advancement in Empirical Investigation into Response Impact. In Coombs, W.T. and Holladay, S.J. (eds) *The Handbook of Crisis Communication*. Malden, MA: Blackwell, pp. 221–242.

Ferrara, V. (2009) Earth Observation and Network of In-Situ Ground Sensors for Disaster Management and Early Warning. In Duncan, K. and Brebbia, C. (eds) *Disaster Management and Human Health Risk: Reducing Risk, Improving Outcomes*. Boston: WIT Press, pp. 3–12.

Fink, S. (2002) *Crisis Management: Planning for the Inevitable*. Lincoln, NE: iUniverse Inc.

Flin, R. et al. (eds) (1997) *Decision Making Under Stress: Emerging Themes and Applications*. Aldershot: Ashgate.

Frandsen, F. and Johansen, W. (2011) The Study of Internal Crisis Communication: Towards an Integrative Framework. *Corporate Communications: An International Journal*, 16(4), pp. 347–361.

Forsberg, T. and Pursiainen, C. (2006) Crisis Decision-Making in Finland: Cognition, Institutions, and Rationality. *Cooperation and Conflict*, 41, pp. 235–260.

Forsberg, T. and Pursiainen, C. (2017) The Psychological Dimension of Russian Foreign Policy: Putin and the Annexation of Crimea. *Global Society*, 31(2).

Gao, S. et al. (2009), Geospatial Web Services and Applications for Infectious Disease Surveillance. In Duncan, K. and Brebbia, C. (eds) *Disaster Management and Human Health Risk: Reducing Risk, Improving Outcomes*. Boston, MA: WIT Press, pp. 13–19.

Gundel, S. (2005) Towards a New Typology of Crises. *Journal of Contingencies and Crisis Management*, 13(3), pp. 106–115.

Haddow, G.D., Bullock, J.A. and Coppola, D.P. (2011) *Introduction to Emergency Management*. Fifth edn. Amsterdam: Elsevier.

Hale, J.E., Dulek, R.E. and Hale, D.P. (2005) Crisis Response Communication Challenges. Building Theory from Qualitative Data. *Journal of Business Communication*, 42(2), pp. 112–134.

Haney, P. (1997) *Organizing for Foreign Policy Crises: Presidents, Advisers and the Management of Decision-Making*. Ann Arbor: University of Michigan Press.

't Hart, P. (1993) Symbols, Rituals and Power: The Lost Dimensions of Crisis Management. *Journal of Contingencies and Crisis Management*, 1, pp. 36–50.

't Hart, P., Rosendahl, U. and Kouzmin, A. (1993) Crisis Decision Making: The Centralization Thesis Revisited. *Administration & Society*, 25, pp. 12–45.

't Hart, P., Stern, E. and Sundelius, B. (1997) *Beyond Groupthink: Political Group Dynamics and Foreign Policy-making*. Ann Arbor: University of Michigan Press.

't Hart, P., Stern, E. and Sundelius, B. (1998) Crisis Management: An Agenda for Research and Training in Europe. *Cooperation and Conflict*, 33, pp. 207–224.

't Hart, P. and Sundelius, B. (2013) Crisis Management Revisited: A New Agenda for Research, Training and Capacity Building Within Europe. *Cooperation and Conflict*, 48(3), pp. 444–461.

Harvard Business School. (2004) *Crisis Management: Master the Skills to Prevent Disasters*. Boston, MA: Harvard Business School Press.

Hermann, C.F. (1963) Some Consequences of Crisis Which Limit the Viability of Organizations. *Administrative Science Quarterly*, 8, pp. 61–82.

Hirschbein, R. (1997) *What if They Gave a Crisis and Nobody Came?: Interpreting International Crises*. Westport, CT: Praeger.

Holsti, O. (1989) Crisis Decision Making. In Tetlock, P.E. et al. (eds) *Behavior, Society and Nuclear War*, Vol. 1. New York: Oxford University Press, pp. 8–84.

Janis, I.L. (1982) *Groupthink: Psychological Studies of Policy Decisions and Fiascoes*. Boston, MA: Houghton Mifflin.

Janis, I.L. (1989) *Crucial Decisions: Leadership in Policymaking and Crisis Management*. New York: Free Press.

Johansen, W., Aggerholm, H. and Frandsen, F. (2012) Entering New Territory: A Study of Internal Crisis Management and Crisis Communication in Organizations. *Public Relations Review*, 38(2), pp. 270–279.

Kahneman, D. and Tversky, A. (1979) Prospect Theory: An Analysis of Decision under Risk. *Econometrica*, 47, pp. 263–291.

Kapucu, N. (2009) Interorganizational Coordination in Complex Environments of Disasters: The Evolution of Intergovernmental Disaster Response Systems. *Journal of Homeland Security and Emergency Management*, 6(1), Article 47, pp. 1–26.

Klinke, A. and Renn, O. (2002) A New Approach to Risk Evaluation and Management: Risk-Based, Precaution-Based, and Discourse-Based Strategies. *Risk Analysis*, 22(6), pp. 1071–1094.

Kouzmin, A. and Jarman, A. (2004) Policy Advice as Crisis: A Political Redefinition of Crisis Management. *International Studies Review*, 6, pp. 182–189.

Lagadec, P. (1997) Learning Processes for Crisis Management in Complex Organizations. *Journal of Contingencies and Crisis Management*, 5(1), pp. 24–31.

Levy, J.S. (1997) Prospect Theory, Rational Choice and International Relations. *International Studies Quarterly*, 41, pp. 87–112.

Maitlis, S. and Christianson, M. (2014) Sensemaking in Organizations: Taking Stock and Moving Forward. *The Academy of Management Annals*, 8(1), pp. 57–125.

Maitlis, S. and Sonenshein, S. (2010) Sensemaking in Crisis and Change: Inspiration and Insights from Weick (1988). *Journal of Management Studies*, 47(3), pp. 551–580.

McDermott, R. (1998) *Risk-Taking in International Politics: Prospect Theory in American Foreign Policy.* Ann Arbor: University of Michigan Press.

McEntire, D.A. (2015) *Disaster Response and Recovery: Strategies and Tactics for Resilience.* Hoboken, NJ: John Wiley & Sons, Inc.

Mintz, A. (1993) The Decision to Attack Iraq. A Noncompensatory Theory of Decision Making. *Journal of Conflict Resolution*, 37(4), pp. 595–618.

Mintz, A. (2004) How Do Leaders Make Decisions?: A Poliheuristic Perspective. *Journal of Conflict Resolution*, 48, pp. 3–13.

Mintz, A. and DeRouen Jr., K. (2010) *Understanding Foreign Policy Decision Making.* Cambridge: Cambridge University Press.

Mintz, A., and Geva, N. (1997) The Poliheuristic Theory of Foreign Policy Decision Making. In Geva, N. and Mintz, A. (eds) *Decisionmaking on War and Peace: The Cognitive-Rational Debate.* Boulder, CO: Lynn Rienner, pp. 81–101.

Morrow, J. D. (1997) A Rational Choice Approach to International Conflict. In Geva, N. and Mintz, A. (eds) *Decisionmaking on War and Peace: The Cognitive-Rational Debate.* Boulder, CO: Lynne Rienner Publishers, pp. 11–31.

Moynihan, D.P. (2009) The Network Governance of Crisis Response: Case Studies of Incident Command Systems. *Journal of Public Administration Research and Theory*, 19: 895–915.

Nicholson, M. (1992) *Rationality and the Analysis of International Conflict.* Cambridge: Cambridge University Press.

Parker, C.F. and Stern, E.K. (2002) Blindsided? September 11 and the Origins of Strategic Surprise. *Political Psychology*, 23(3), pp. 601–630.

Popper, K. (1994) *The Myth of Framework: In Defence of Science and Rationality.* London: Routledge.

Post, J.M. (2004) *Leaders and Their Followers in a Dangerous World: The Psychology of Political Behaviour.* Ithaca, New York: Cornell University Press.

Preston, T. and 't Hart, P. (2011) Understanding and Evaluating Bureaucratic Politics: The Nexus between Political Leaders and Advisory Systems. In Carlsnaes, W., and Guzzini, S. (eds) *Foreign Policy Analysis I-V, Volume III.* Sage Library of International Relations, Los Angeles: Sage, pp. 227–278.

Pursiainen, C. (2007) Why Early Warning Sometimes Fails: The Case of Civil Protection. Working Paper. Stockholm: Nordregio.

Regester, M. and Larkin, J. (2008) *Risk Issues and Crisis Management in Public Relations: A Casebook of Best Practice.* Fourth edn. London: Kogan Page.

Roberts, J. (1988) *Decision-Making During International Crises.* New York: St Martin's Press.

Robertson, A. (2001) Mediated Threats. In J. Erikson (ed.) *Threat Politics: New Perspectives on Security, Risk and Crisis Management.* Aldershot: Ashgate.

Rosenthal, U., Charles, M.T. and 't Hart, P. (1998) *Coping with Crises: The Management of Disasters, Riots, and Terrorism.* Springfield, IL: Charles C. Thomas.

Rosenthal, U., 't Hart, P. and Kouzmin (1991) The Bureau-Politics of Crisis Management. *Public Administration*, 69(2), pp. 211–233.

Ruck, K. and Welch, M. (2012) Valuing Internal Communication: Management and Employee Perspectives. *Public Relations Review*, 38(2), pp. 294–302.

Scholtens, A., Jorritsma, J. and Helsloot, I. (2014) On the Need for a Paradigm Shift in the Dutch Command and Information System for the Acute Phase of Disasters. *Journal of Contingencies and Crisis Management*, 22(1), pp. 39–51.

Seeger, M.W. (2006) Best Practices in Crisis Communication: An Expert Panel Process. *Journal of Applied Communication Research*, 34(3), pp. 232–244.

Sellnow, T.L., Seeger, M.W. and Sellnow, T.L. (2013) *Foundations in Communication Theory: Theorizing Crisis Communication, Vol. 1*. Somerset, NJ: Wiley-Blackwell.

Simon, H.A. (1947) *Administrative Behavior*. New York: Free Press.

Smith, S. (1980) Allison and the Cuban Missile Crisis: A Review of the Bureaucratic Politics Model of Foreign Policy Decision-Making. *Millennium: Journal of International Relations*, 9(1), pp. 21–40.

Stein, J. and Welch, D.A. (1997) Rational and Psychological Approaches to the Study of International Conflict: Comparative Strengths and Weaknesses. In Geva, N. and Mintz, A. (eds), *Decisionmaking on War and Peace: The Cognitive-Rational Debate*. London: Lynne Rienner Publishers Inc., pp. 51–77.

Stern, E. (1999) *Crisis Decision-making: A Cognitive Institutional Approach*. Stockholm Studies in Politics 66. Stockholm: University of Stockholm, Department of Political Science.

Stern, E. (2013) Preparing: The Sixth Task of Crisis Leadership. *Journal of Leadership Studies*, 7(3), pp. 51–56.

Stern, E. and Sundelius, B. (2002) Crisis Management Europe: An Integrated Regional Research and Training Program. *International Studies Perspective*, 3, pp. 71–88.

Strandberg, J.M. and Vigsø, O. (2016) Internal Crisis Communication. *Corporate Communications: An International Journal*, 21(1), pp. 89–102.

Sylvan, D. (1998) *Problem Representation in Foreign Policy Decision Making*. Cambridge: Cambridge University Press.

UK Cabinet Office. (2013) *Emergency Response and Recovery. Non-statutory Guidance Accompanying the Civil Contingencies Act 2004*. Civil Contingencies Secretariat. Available at: https://www.gov.uk/government/uploads/system/uploads/attachment_da ta/file/253488/Emergency_Response_and_Recovery_5th_edition_October_2013.pdf

UNISDR. (2009) UNISDR Terminology on Disaster Risk Reduction, United Nations International Strategy for Disaster Reduction (UNISDR). Geneva, Switzerland, May 2009. [Online]. Available at: http://www.unisdr.org/we/inform/terminology

Vetschera, H. (2005) Early Warning in the Yugoslav Crisis and the Development of Instruments – A European Perspective. *Dialogue + Cooperation*, 3, pp. 57–70.

Vogler, J. (1989) Perspectives on the Foreign Policy Systems: Psychological Approaches. In Clarke, M. and White, B. (eds) *Understanding Foreign Policy: The Foreign Policy Systems Approach*. Cheltenham: Edward Elgar, pp. 135–162.

Wæver, O. (1995) Securitization and Desecuritization. In Lipschutz, R. (ed.) *On Security*. New York: Columbia University Press, pp. 46–86.

Watters, J. (2014) *Disaster Recovery, Crisis Response, and Business Continuity. A Management Desk Reference*. New York: Apress.

Weick, K.E. (1988) Enacted Sensemaking in Crisis Situations. *Journal of Management Studies*, 25, pp. 305–317.

Weick, K.E. (1995) *Sensemaking in Organizations*. Thousand Oaks, CA: Sage.

Welch, D.A. (1992) The Organizational Process and Bureaucratic Politics Paradigms: Retrospect and Prospect. *International Security*, 17(2), pp. 112–146.

Wimelius, M.E. and Engberg, J. (2015) Crisis Management through Network Coordination: Experiences of Swedish Civil Defence Directors. *Journal of Contingencies and Crisis Management*, 23(3), pp. 129–137.

6 Recovery

Merriam-Webster refers to recovery as "the act or process of returning to a normal state after a period of difficulty". Obviously, this very generic definition can be further refined and tailored depending on the issue area. Focusing on disaster management, for instance, UNISDR (2009) defines recovery as the "restoration, and improvement where appropriate, of facilities, livelihoods and living conditions of disaster-affected communities, including efforts to reduce disaster risk factors". Further, it is mentioned that the recovery task of rehabilitation and reconstruction "begins soon after the emergency phase has ended, and should be based on pre-existing strategies and policies that facilitate clear institutional responsibilities for recovery action and enable public participation". The various definitions suggested by national and international organisations (cf. CIPedia, n.d.) are basically only slight variations of the ones above.

According to a comprehensive and rather recent literature review on crisis management – divided in this context into mitigation, preparedness, response, and recovery – the latter is the subject that has received the least treatment, constituting little more than 3 per cent of all literature (Galindo and Batta, 2013; cf. Altay and Green, 2006). There is, however, a burgeoning body of resilience literature, which has a lot in common with recovery. It can be argued that while resilience is not synonymous with recovery, the latter is perhaps the former's central concept. This chapter is therefore constructed taking into account both traditional crisis recovery literature and more recent resilience literature. All in all, the literature on recovery remains rather sparse.

The chapter is divided into seven sections. The first section can be understood as an introduction of sorts, focusing on the more traditional genre of recovery literature that pays greater attention to planning recovery than it does to its actual post-response phase. The second section then focuses on the concept of resilience, particularly from the recovery perspective. This is followed by five brief sections, each dedicated to one thematic area of recovery or resilience: societal, organisational, technological, economic and psychological.

6.1 Recovery as planning

Most of the explicit recovery literature – indeed, the majority of those books that include the concept in their title (e.g. McEntire, 2015; Watters, 2014; Varghese, 2002) – does not really have much to say about recovery, concentrating instead on preparedness and response issues that facilitate recovery later on. Haddow, Bullock and Coppola (2011, pp. 257, 258), focusing on disasters, explicitly emphasise that communities have to conduct long-term recovery planning "before the next disaster strikes". The main goals are to identify the most vulnerable areas of the community and to anticipate the need for internal and external assistance needs, and consequently to enhance preparedness. In essence, what they propose is conducting activities related to risk assessment, prevention and mitigation, and preparedness, already discussed in this book as separate crisis management phases. This planning should establish the financial arrangements for subsidising the rehabilitation of public infrastructures for local governments, as well as prior agreements with the private sector to immediately commence recovery works (Ishiwatari, 2014). Austina, Fisher Liub and Jin (2014) have also noted that the mere symbolism of repairing damage and rebuilding sends an important message to the media and the public, and planning recovery in advance enhances this symbolism.

The overwhelming body of recovery literature relevant to the current book's context is on disaster recovery, often focusing on large-scale natural catastrophes and disasters. The bulk of this consists of experience-based guidelines and best practices, often country-based. Looking at community recovery planning, the UK Cabinet Office (2013) guidelines provide a particularly good starting point. Recovery is defined in this context as "the process of rebuilding, restoring and rehabilitating the community following an emergency", including four interlinked categories of impact, namely humanitarian, economic, infrastructural and environmental. The guidelines emphasise that recovery from a major emergency is a complex, costly and long-running process that involves many more agencies and participants than the response phase. That is why, during the recovery phase, the structures, processes and relationships that underpin it are harder to get right than in the response phase. It is therefore essential that the recovery structures and processes are well-planned in advance, and even tested. Recovery planning and management arrangements should be supported by training programmes and multi-agency exercises, which ensure that the agencies and groups involved in the future recovery process are properly prepared for their roles.

Moreover, the guidelines suggest that one should not delay the start of recovery, contrary to the UNISDR definition above, and that this should begin at the earliest opportunity and run in tandem with the response. In practical terms, the guidelines furthermore recommend a formal handover process from those responsible for response to those responsible for recovery. Recovery then continues until the disruption has been rectified, the demands on services have returned to normal levels, and the needs of those affected have been met. Hence the recovery phase may last for months, years or even decades.

In any case, effective recovery requires planning. The management of recovery is most effective when conducted at the local level with the active participation of the affected community and a strong reliance on local capacities and expertise. The private sector, the voluntary sector and the wider community should all play a crucial role, not just the authorities. The after-the-crisis recovery activities should start with an impact assessment. Part of the recovery strategy is to set targets and milestones. The activities should then be documented by means of a reporting framework for recovery.

More academic but still heavily normative literature basically confirms the practical guidelines and rules of conduct outlined above. To this end, Davis and Alexander (2016) have listed several principles that should be taken into account in recovery from disaster. First, similarly to the above guidelines, they emphasise the importance of community participation. Disaster survivors need to be closely involved in decisions that affect them. International and national aid organisations should not take over the recovery operation but, rather, their role is to empower community efforts and citizens to regain control over their own lives. Second, recovery should be understood as a multi-sectoral process, combining psychological, environmental, economic, physical, political and administrative elements. Third, the role of national governments is to coordinate recovery actions by harmonising the local level procedures and ensuring that there is strong leadership in recovery. Fourth, and perhaps most importantly, in all recovery actions a key factor should be to improve safety to reduce the likelihood of future disasters.

While community participation is emphasised in almost all of the literature on recovery, the issue of just what exactly the role of local ownership (decentralisation) and that of national government (centralisation) should be is somewhat more complex or contested. Clearly, the normative lessons and recommendations reflect the political and administrative cultures of the network of crisis management actors, and thus vary from country to country. A study on Chinese recovery efforts after a serious earthquake (Chang et al., 2009, pp. 319, 320) suggests that a mixed system of centralised and decentralised recovery processes is the most functional. The authors propose that the lessons learned include the need to build a strong organisational structure to deal with recovery, while at the same time keeping the local social communities involved. But, according to the study, it is "central policy planning with a decentralized mechanism to ensure decision making and involvement of all players and implementation of a recovery plan" that will make the recovery run smoothly.

The last point in Davis and Alexander's (2016) above-mentioned normative list seems to be common to most of the literature. In disaster recovery literature in particular – but applicable to any field – criticism is often expressed of the suggestion that post-disaster recovery should be framed in terms of a return to the pre-disaster conditions. The critique centres on the fact that recovery "too often means rebuilding pre-existing conditions of disaster risk, thus preparing the ground for future disaster" (Shaw, 2014, p. 5). Instead, the challenge lies in realising that recovery is an opportunity for building a resilient society, characterised by the notion of rebuilding better than before at a reasonable pace; or local communities may use a disaster as an opportunity to regenerate an area, in which case this regeneration phase may overlap with the recovery phase (UK Cabinet Office, 2013). This then leads to the argument that while there is always time pressure to rebuild critical structures and systems, ultimately "recovery is a balance between speed and quality" (Shaw, 2014, p. 5; cf. UNISDR, 2009).

While each crisis is idiosyncratic, in disaster recovery studies there are several general themes and challenges that are largely unrelated to the nature of disaster. One of them is coordination between the different actors involved in recovery (e.g. Kapucu, 2014). Some studies (e.g. Thiruppugazh, 2014; Wen et al., 2014) propose that recovery plans and the actual recovery in large-scale disasters should not rely on the existing governance and decision-making models but should benefit from the creation of an extraordinary governance mechanism. This kind of extra bureaucracy should bring together various agencies and departments for collective decision-making, thereby speeding up the decision-making process and working as a single window for donors and lenders, which will reduce their burden of having to interact with multiple government agencies.

Most often, the focus in recovery literature is on local communities, but even if the focus is on the businesses, the themes and recommendations are basically the same: namely, plan for recovery! Runyan (2006), for instance, has identified a lack of planning by small businesses vis-à-vis external disasters, such as natural catastrophes. However, the issues that require advance planning differ in part from the community recovery perspective. For Runyan, small businesses face such challenges as vulnerability to cash flow interruption, lack of access to capital for recovery and serious infrastructure problems impeding recovery. Watters (2014, pp. 215–219), who also focuses on business recovery, mainly in terms of production and data recovery, similarly emphasises that naturally "it will be easier to return to normal if you have planned for it in advance". For him, this entails preparing a recovery action plan for various types of threat scenarios, as well as testing the plan. The key concept is a business continuity plan (BCA) (ibid., pp. 49–56). Watters proposes starting by identifying which activities are critical for one's business, similar to the way in which one approaches risk assessment. He calls this process business impact analysis (BIA) (ibid., pp. 35–48), and it includes creating a structured questionnaire to this effect, completing it and then analysing the results. The results should indicate which systems are so critical that their recovery should be prioritised, and how quickly this has to be done. In so doing, one should take into account legal, regulatory, customer, financial and reputational perspectives. The type of impact can be formally rated and weighted in order to establish priorities, and one should be able to establish thresholds as to the expected time of recovery. One can also establish thresholds for expected loss values in monetary terms. In order to avoid confusion and contradictory assessments, Watters proposes a top-down, executive-led BIA approach, whereby he or she interviews the department or unit heads and other stakeholders, and then determines the priorities. The result is a typical risk assessment on the likelihood of the critical functions not being recovered in the expected time and within acceptable costs. Similarly, as will be discussed below in terms of resilience, Watters (ibid., pp. 84–98) then identifies the business recovery options or strategies, in the same way as one prepares risk treatment strategies. As the focus is on business continuity, Watters lists the following key strategies: work transfer to another location; displacement of non-critical functions or people to provide room for critical staff; doubling up, that is, working in shifts; splitting the workload between two sites; offshoring; work area recovery by providing alternative workspaces to operate critical functions; and mutual aid based on pre-agreed contracts. One usually needs a blended approach towards these options, which together make up the overall recovery strategy.

6.2 Resilience as recovery

Although the concept of resilience has deep roots in many disciplines, in its contemporary meaning there are grounds for tracing it back to the ecological debates of the early 1970s (Holling, 1973). The concept became popularised in unofficial policy and scientific analyses in the mid-2000s in the context of crisis and disaster management as well as critical infrastructure (Pursiainen and Gattinesi, 2014). Today, resilience is the ultimate catchword in safety and security debates.

In the early debates on resilience, there was often criticism of official government positions that were perceived to be overly protection-oriented, on the basis that complete protection, of critical infrastructure, for instance, can never be guaranteed. Moreover, achieving the desired guaranteed level of protection is normally not cost-effective in relation to the actual threats. A small increment in the level of protection might introduce hefty additional costs, and hence alternative approaches need to be considered. Thus, as De Bruijne and Van Eeten (2007) put it, a fringe benefit from a more resilience-based preparation approach is that these measures are substantially less expensive than investments in specific infrastructure or other upgrades to avoid certain risk scenarios, which may or may not occur.

Expressing the same idea in more idiomatic terms, Landstedt and Holmström (2007) proposed that the characteristics of protection and resilience could be likened to the characteristics of a rigid stick and a flexible one. The former is harder to bend, but will snap under severe pressure and cannot be repaired. By contrast, the flexible stick is easy to bend, always regains its shape and is hard to break. Therefore, they argue, focusing on protection alone may provide a false

132 *Recovery*

sense of security, which can turn out to be disastrous, as has been proved many times in the past. For Landstedt and Holmström, resilience is a much wider concept than protection, encompassing not only protection itself, but also prevention, training, education, research, deterrence, risk-based mitigation, response, recovery and longer-term restoration.

Numerous scholarly, national and international definitions of resilience exist, depending on the field of interest (cf. CIPedia, n.d.). An appropriate and sufficiently generic definition for the current purpose is again provided by the UNISDR (2009):

> The ability of a system, community or society exposed to hazards to resist, absorb, accommodate to and recover from the effects of a hazard in a timely and efficient manner, including through the preservation and restoration of its essential basic structures and functions.

Hence, the basic idea is that any system, be it a country, municipality, organisation or company, should be resilient in such a way that when it is no longer able to resist stress, it nonetheless adapts to the ensuing crisis but, more importantly, its basic characteristics and functions, such as service level, will rapidly be restored.

It is notable that there is a certain temporal dimension to resilience in this definition, and recovery is only one of its 'phases'. In fact, the afore-mentioned UNISDR definition of resilience covers most of the same issues that have been discussed in the current book, including phases before a crisis ("resist", "absorb"), during a crisis ("accommodate") and after a crisis ("recover"). In the approach developed by the US Department of Homeland Security (HSSAI, 2009; cf. Nieuwenhuijs, Luiijf and Klaver, 2008), three interrelated and reinforcing resilience phases are identified, namely, resistance, absorption and restoration. The same thinking is also expressed in official policies, such as the US cybersecurity strategy (Executive Order, 2013), which is based on the idea of five functions: identify, protect, detect, respond and recover. In other words, resilience is often approached through crisis management parlance, being a near synonym (e.g. Pursiainen et al., 2016; Kozine and Andersen, 2015; Petit, Wallace and Phillips, 2014; Petit et al., 2013). In the context of the current chapter, resilience is discussed in terms of a system's restoration capability, which is basically the same concept as recovery. The term rapidity is also sometimes used.

Following HSSAI (2009) terminology, if we combine the concepts of resistance, absorption and restoration in the same diagram, we get Figure 6.1, illustrating the basic idea of the so-called resilience or performance loss triangle (cf. Chang et al., 2014; Wang and Blackmore, 2009; Bruneau et al., 2003; McDaniels et al., 2007).

Figure 6.1 depicts three critical systems called A, B and C, with the vertical axis representing the service level and the horizontal axis representing time. Let us consider that the hazard causing the event is similar for all three

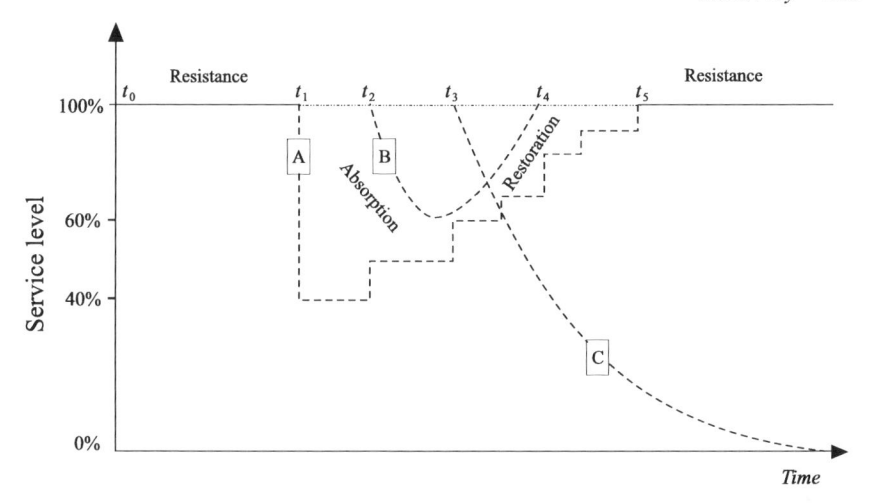

Figure 6.1 Resilience of three systems: A, B and C

systems. In Figure 6.1, t_0 indicates the time when the stress exerted against the system starts. As we can see, the systems differ in terms of their resistance to this stress. The function performance curve A illustrates a system that is performing worse in terms of resistance (prevention), as it fails as early as t_1. When it breaks down, its service level also drops directly downwards, reaching a rather low point of 40 per cent, which means that its absorptive (response) capacity is no more than moderate. Its restoration (recovery) capacity is also rather moderate as it is restored only by degrees and not until the rather late t_5. Curve B, in turn, illustrates a system that is more resistant and that withstands the stress until t_2 ($>t_1$). It is also more absorptive than A as it does not plummet so steeply and dip so low in terms of service level. Furthermore, it is able to restore rather quickly as it is back by t_4 ($<t_5$). Curve C, finally, illustrates a system that is highly resistant, indeed until t_3 ($>t_2>t_1$), but when the damage hits, it leads to a total long-term or permanent function failure.

In a sense, Figure 6.1 illustrates the different resilience strategies through which organisations deal with hazards (cf. Gibson and Tarrant, 2010, p. 11). The resilience strategy reflects the level of investment in resilience, or more specifically in its different phases, the attention paid to related technological and organisational solutions, and the level of strategic thinking in an organisation in general. If system B's resilience curve most clearly resembles a triangle, the fundamental idea is that reducing the triangle in all its dimensions and moving it to the right would increase resilience.

How then to measure resilience? Moteff (2012) has proposed examining the level of performance, that is, whether the system was put out of operation completely or not. Consequently, reducing the cases of total loss of performance would then increase resilience. He adds another measure, however, which is the amount of time (or money/losses) it would take for an infrastructure or a function to be fully restored to normal operations; naturally, the quicker the recovery, the more resilient the system. Similarly, Ford et al. (2012) have noted that simply considering performance outputs may not be enough. Another parameter to consider is the capacity of the system to recover from subsequent failures or attacks. The same issue has been discussed in terms of recoverability (Barker, Ramirez-Marquez and Rocco, 2013), which is the speed or rapidity at which a system recovers.

What this boils down to is establishing the correlation between the planned preparedness for recovery and the actual recovery. For analytical purposes, and to measure a system's capacity to recover, it is wise to keep the facilitating plans and preparations for recovery strictly separate from the actual recovery. The actual recovery should therefore be measured based solely on historical incident data, measuring the real recovery ability of a system.

6.3 Societal recovery

Arguably, most of the crisis recovery literature focuses on disaster recovery. The line between response and recovery is, however, a line in the water. Somewhere in between is what is sometimes called "moving beyond the immediate needs". McEntire (2015, pp. 236–271) discusses this in terms of damage assessment. He differentiates between a rapid assessment, which is the initial survey of damages, and a preliminary assessment, which is a much more detailed procedure. This is followed by a technical assessment, which is conducted to determine the cost of damages and method of recovering or rebuilding the damaged structures and systems. After that comes the actual recovery phase. In the context of the US emergency management system, McEntire (2015, pp. 272–306) discusses this mainly in terms of assistance programmes that can help communities and individuals to recover, including loans, grants, services and similar forms of public assistance.

In the very same spirit, Haddow, Bullock and Coppola (2011, pp. 229–262) note that the recovery phase is not easily classified as it overlaps with that of response. It often begins in the initial hours of the disaster and can continue for months or, in some cases, for years. They also emphasise the need for technical and financial support, such as housing programmes, for the communities affected, as well as good coordination between different levels of government and non-governmental actors.

In terms of more generic resilience discourse, one speaks about societal, social or community resilience. In this respect, there are numerous examples of good practices exhibited by resilient communities. The focus in the societal resilience debates is on the local community's problems when it faces crises, emergencies or disasters. As communities are usually victims of crises not of their own making but caused, for instance, by natural disasters or critical infrastructure service disruption, the question is usually one of adaptive (response) and restoration (recovery) capacities in the face of these disturbances. A related concept is socio-ecological resilience (Francis and Bekera, 2014, pp. 92, 94, 95, 100–102), a system of systems that depends on a wide range of factors stemming from the linkages between human societies and ecosystems. Its focus is very much on the adaptive and self-organising response capacity of a community rather than on recovery. However, recovery is indirectly involved as the concept proposes features such as diversity, efficiency, adaptability and cohesion, which would reduce the vulnerability of social and engineered systems in the event of unforeseen and unanticipated disruptions. Obviously, this kind of system is expected not only to adapt to, but also to recover from crises smoothly.

There are no definitive, agreed-upon metrics to evaluate societal or community resilience. Moreover, many of the societal resilience approaches are highly generic, and thus difficult to operationalise. While quite a few efforts to develop societal resilience indicators and indices exist (e.g. LEDDRA Project, 2014; Boon et al., 2012; Sherrieb, Norris and Galea, 2010; McAslan, 2010; Cutter, Burton and Emrich, 2010; Cutter et al., 2008a; Cutter et al., 2008b; Norris et al., 2008; Flint and Luloff, 2007; Flint and Luloff, 2005; Cumming et al., 2005; Klein, Nicholls and Thomall, 2003; Bruneau et al., 2003), they often only list socio-economic or institutional–political indicators at a very general level.

For instance, Cutter, Burton and Emrich (2010) present a set of indicators for measuring baseline levels of community resilience (and therefore also recovery capacity) in the US. The authors suggest that social resilience could be measured by analysing such indicators as educational equity, age, transportation access, communication capacity, language competency, special needs, health coverage, place attachment, political engagement, social capital in terms of religion, social capital in terms of civic involvement and advocacy, and innovation. Similarly, LEDDRA Project (2014; cf. Wilson, 2012, pp. 4–47) suggests that social resilience could be best measured by evaluating a community's identity, cohesiveness and trust, societal relationships, contentment with life, conflicts, communication between stakeholder groups, power, political structures, engagement of young people, responses to and opportunities for influencing change, learning and knowledge, knowledge utility and transfer, learning from experience, participation in decision-making, engagement of community resources and stakeholder agency.

At best, in terms of concreteness, the literature on societal resilience suggests indicators that reflect the emergency management and self-assistance capacities of the community. While the concept has its shortcomings, it is most useful from the perspective that its emphasis is on the holistic picture of a society's resilience, including recovery capacity, in times of disasters. In more recent literature, one can find efforts to consider the linkage between infrastructures and social systems (Chang et al., 2014), arguing that there is a need to link physical systems and human communities in order to measure and enhance societal resilience.

It is also noteworthy in societal recovery or resilience debates, as in the crisis and resilience literature in general, that a return to pre-event conditions may be short-sighted if these conditions are themselves in need of improvement. A crisis duly offers a series of unique and valuable opportunities to improve on the status quo. Hence, one should capitalise on these opportunities to advance the conditions for becoming more resilient (IoM, 2015).

6.4 Organisational recovery

McManus (2008, p. 82), among others, suggests a definition of organisational resilience that is based on different temporal phases. For him, organisational resilience is "a function of an organization's overall situation awareness, management of keystone vulnerabilities and adaptive capacity in a complex, dynamic and interconnected environment". If we were to consider the same thing from a business point of view, recovery or resilience would be closely related to the concept of business continuity (e.g. Watters, 2014), as revealed by the title of one of the cornerstones of this discourse, namely Sheffi's (2005) monograph *The Resilient Enterprise*. Keeping the business running is normally the key driver for any enterprise, not to mention infrastructure operators. The failure of a company to deliver a service could quickly lead to financial disaster for the owner, regardless of the impact on society, which might be able to rely on an alternative service provider. Basically, Business Continuity Management, as it is called, is based on the same issues presented in this book in terms of the crisis management cycle, starting from risk assessment and ending with recovery (e.g. Gibb and Buchanan, 2006). Thus, it is largely a question of normal risk and the crisis management of an enterprise.

But how does an organisation know whether it is resilient or not? Indeed, there is a growing body of literature that is specifically aimed at developing indicators to measure an organisation's resilience (e.g. AIIC, 2016; Pursiainen et al., 2016; Hosseini, Barker and Ramirez-Marquez, 2016; Labaka, Hernates and Sarriegi, 2015; Prior, 2015; Petit, Wallace and Phillips, 2014; Petit et al., 2013; Linkov et al., 2013; Gibson and Tarrant, 2010; Stephenson, 2011; Kahan, Allen and George, 2009). Furthermore, there is also a number of national and international standards in this field (ISO, 2014a-c; ISO, 2011; cf. ISO, 2007; ISO/IEC, 2005; ISO, 2004; ISO, 2000; ANSI/ASIS, 2012; ANSI/ASIS, 2009; BS, 2014). It is notable that the first resilience standards are specifically related to organisational resilience. Thus, the ISO 28002 standard for resilience in the supply chain was approved in 2011, based on the US ANSI/ASIS organisational resilience standard.

The focus of this literature is primarily on organisations that own and manage critical infrastructure facilities. In more general terms, this literature is very much focused on the challenges of the supply chain in the context of potential infrastructure failures (Briano, Caballini and Revetria, 2009), and the first resilience standards are related to this issue, in fact. The purpose is to measure the ability of an organisation to withstand a disturbance in its critical infrastructure facilities and to maintain or quickly regain its functionality. In practice, this is realised in a self-auditing manner for the most part, motivated by self-interested profit-seeking in terms of business continuity, although public good considerations might also be taken into account, at least for the sake of possible reputational costs.

In order to be resilient, organisations must take into account such factors as strong and flexible leadership, an awareness and understanding of their operating environment, their ability to adapt in response to rapid change and so forth. Yet, while at first sight this appears to be a rather straightforward process, and as such suitable for universal standardisation, it becomes more complicated due to the fact that social and cultural idiosyncrasies and differences must also be considered (Lee, Vargo and Seville, 2013). An organisation that is resilient in one environment might not be resilient in another. Indicators such as those that are particularly needed in response and recovery from a crisis, including innovativeness, creativity and the improvisation skills of the organisation's leadership, are also strongly advanced in this literature (e.g. Stephenson, 2011), but are nevertheless rather difficult to measure, except post factum.

6.5 Technological recovery

Technological (or technical or engineering) resilience naturally starts from an engineering point of view, and in the current literature it is mostly applied to the field of critical infrastructure. While technological resilience includes elements of organisational resilience, and these two domains are mutually dependent in many cases, the main difference is that resilience is achieved by technological rather than organisational solutions. Naturally, the literature on technological resilience (e.g. Cimellaro, 2016) differs in essence from that on organisational resilience, focusing instead on engineering systems or subsystems that would enhance recovery. The main actors in the context of this domain of resilience are critical infrastructure operators, that is, those very facilities that produce critical services, such as power production or drinking water management. The regulators' role is to set the technical standards and to ensure that they are being followed. Furthermore, in most cases, the in-house technological or engineering capacities and capabilities of a service producer are inadequate, and one has to rely on manufacturers or vendors for resilience-related technological solutions.

There is no officially approved definition of technological resilience in terms of international standards. However, a certain level of consensus has been emerging in the related literature. From the standard definition of resilience, one can already derive the main elements of technological resilience. If a resilient infrastructure can be defined as a component, system or facility that is able to withstand damage or disruption, and which, if affected, can be readily and cost-effectively restored, then there are two key technological concepts in resilience that should be demanded from any resilient critical infrastructure: resistance and restoration capacity. Every engineering solution naturally has its own specific features, but this rather minimalist definition provides us with quite a straightforward understanding of the general attributes or elements that we could measure when talking about resilient infrastructure, especially from a technological perspective.

One can also find more detailed typologies and indicators in the literature on technological resilience (e.g. Pursiainen et al., 2016; Hosseini, Barker and Ramirez-Marquez, 2016; Sahebjamniaa, Torabia and Mansourib, 2015; Labaka, Hernates and Sarriegi, 2015; Prior, 2015; Petit, Wallace and Phillips, 2014; Petit et al., 2013; Vlacheas et al., 2013; Linkov et al., 2013; Sterbenz et al., 2011; Youn, Hu and Wang, 2011; McAslan, 2010b; Kahan, Allen and George, 2009). In fact, as early as 2003 Bruneau et al. provided a more detailed typology of earthquake resilience that could be applied to critical resilience as well. This typology included four levels: robustness, redundancy, resourcefulness and rapidity (of recovery), with only the latter being relevant from a recovery point of view. This typology, in turn, has been repeated with some variations in most definitions of resilience (cf. Flynn, 2008). Rapid recovery in these definitions refers to the capacity to get things back to normal as quickly as possible after a disaster.

There are several elements or characteristics that can enhance recovery. This may include, for instance, unplanned maintenance, how easy it is to restart the system, the autonomy of the system from other systems that may be down, etc. These attributes in turn will be affected by several other indicators. Thus unplanned maintenance, for instance, depends on such factors as whether maintenance personnel and spare parts are available, or whether the system has been designed to be resilient from an unplanned maintenance point of view. The latter factor is often called maintainability. Just to illustrate, it is clear that if a technological system is already designed to be resilient, it should, as far as possible, be based on modules; technological failure in one part of the system could then be quickly repaired by removing the failed module and replacing it with a new one. This kind of system then would be more maintainable than systems without this characteristic, and thus could be restored more quickly in the case of a technological failure. However, this kind of recovery ability is not stable even within the same technological system. While a technological failure may be relatively quickly repaired under the right conditions, it also depends on many situational factors, such as the type of failure and its wider societal dimensions, the weather, the season and even the time of day. In any case, the task is to identify the elements that are important for recovery, measure them on some scale, aggregating the different elements into a recovery index, and in this process enhance recovery ability (Pursiainen et al., 2016).

6.6 Economic recovery

A field of study emerged in the early 2000s focusing on economic resilience in particular. Obviously the issue here is, broadly speaking, to study the economic consequences of either the existence or absence of resilience, with the aim of proposing and encouraging resilience measures, including economic or financial recovery. A review of the literature on economics indicates that most in-depth studies of economic resilience emerged first in the area of ecological economics, with resilience sometimes characterised in a typical Darwinian sense: systems, like species, that cannot adapt are unlikely to survive. This literature specifically emphasised the need to take into account the lack of predictability and the disequilibrium nature of dynamics, as well as the chaos theory axiom that even small changes can be magnified into system failures.

When framed in more economic terms, the main focus in this literature is on the dynamics of technological innovation and the need for adaptive behaviour in businesses and markets (Rose, 2009, pp. 7, 8). Consequently, issues such as the extent of regional economic diversification; the ability to substitute and conserve necessary inputs; the commercial and industrial capacity to improvise; and the time needed to regain capacity or lost revenues are often emphasised in this field of study (O'Rourke, 2007).

Rose (2009) defines economic resilience by differentiating between two types of resilience: static economic resilience is "the ability of an entity or system to maintain function (e.g. continue producing) when shocked", whereas dynamic economic resilience is "the speed at which an entity or system recovers from a severe shock to achieve a desired state". Rose and Krausman (2013) have further divided the 'economic' type into three levels: microeconomic (individual business or household), mesoeconomic (individual industry or market) and macroeconomic (combination of all economic entities). As they note, the latter overlaps considerably with the community or societal resilience focus. The idea then is that one develops economic resilience indicators for all of the aforementioned three levels. Taken together, they would then constitute an overall index for characterising economic resilience.

6.7 Psychological recovery

One also speaks about personal resilience (Bearse and Coss, 2014) or psychological resilience (Rodriguez-Llanes, Vos and Guha-Sapir, 2013). Psychological resilience concerns the capacity for recovery at a personal level. This is closely related to the more traditional psychological recovery debate and research focusing on post-traumatic stress disorder (PTSD). The diagnostic criteria for PTSD, stipulated in the International Statistical Classification of Diseases and Related Health Problems 10 (ICD-10, n.d.), advances the following diagnostic criteria for PTSD. First, a person is exposed to a stressful event or situation (either short- or long-lasting) of an exceptionally threatening or catastrophic nature, which is likely to cause pervasive distress in almost anyone. Second, the person experiences persistent recall, or relives the stressor through intrusive flashbacks, vivid memories, recurring dreams or by experiencing distress when exposed to circumstances resembling or associated with the stressor. Third, the person seeks actual or preferred avoidance of circumstances resembling or associated with the stressor, not present before exposure to the stressor. Fourth, the person experiences an inability to recall, either partially or completely, some important aspects of the period of exposure to the stressor, or alternatively, the person experiences persistent symptoms of increased psychological sensitivity and arousal, not present before exposure to the stressor. Fifth, the person experiences symptoms such as difficulty falling or staying asleep, irritability or outbursts of anger, difficulty concentrating, hyper-vigilance and an exaggerated startle response.

PTSD can result from a variety of situations, not only for those who are seen as the victims of a crisis, but also for the rescuers who see the victims. Sifaki-Pistolla et al. (2017), for instance, have identified serious PTSD among rescue workers operating during the European refugee crisis. The chain of cause and effect can sometimes be very complicated. For instance, Kanellopoulos et al. (2014) have shown how a financial crisis leaving both parents unemployed can be reflected in their children being exposed to bullying at school, and result in them having significantly higher PTSD scores on average.

Hence, psychological resilience discourse focuses more on the question of how to measure whether a person is psychologically resilient or not, especially to PTSD. Rodriguez-Llanes, Vos and Guha-Sapir (2013), after conducting a considerable literature review and carrying out tests, found that there are numerous factors that affect psychological resilience. Those indicators which mostly cause low resilience to disasters are lack of social support, female gender, prior traumas, resource loss, human loss and poor physical or mental health. It is then proposed that policies improving these conditions or targeting the most vulnerable groups in advance might be effective in increasing psychological resilience. It has also been found that, in general, a higher level of education is predictive of a resilient psychological outcome, and sometimes a higher income, whereas the effect of age on psychological resilience remains unclear.

References

AIIC. (2016) *Guidelines for Critical Infrastructures Resilience Evaluation*. Italian Association of Critical Infrastructures' Experts (AIIC).
Altay, N. and Green, W.G. (2006) Or/Ms Research in Disaster Operations Management. *European Journal of Operational Research*, 175(1), pp. 475–493.
ANSI/ASIS. (2009) *Organizational Resilience: Security, Preparedness, and Continuity Management Systems – Requirements with Guidance for Use*. ANSI/ASIS.SPC.1.
ANSI/ASIS. (2012) *Maturity Model for the Phased Implementation of the Organizational Resilience Management System*. ANSI/ASIS.SPC.4.
Austina, L.L, Fisher Liub, B. and Jin, Y. (2014) Examining Signs of Recovery: How Senior Crisis Communicators Define Organizational Crisis Recovery. *Public Relations Review*, 40, pp. 844–846.
Barker, K., Ramirez-Marquez, J.E. and Rocco, C.M. (2013) Resilience-Based Network Component Importance Measures. *Reliability Engineering & System Safety*, 117, pp. 89–97.

Bearse, R. and Coss, A. (2014) The Return on Investing in Personal Resilience. *The CIP Report, Center for Infrastructure Protection and Homeland Security*, 12(7), pp. 21–24.

Boon, H.J. et al. (2012) Bronfenbrenner's Bioecological Theory for Modelling Community Resilience to Natural Disasters. *Natural Hazards*, 60(2), pp. 381–408.

Briano, E., Caballini, C. and Revetria, R. (2009) Literature Review about Supply Chain Vulnerability and Resiliency. Proceedings of the 8th WSEAS International Conference on System Science and Simulation in Engineering.

Bruneau, M. et al. (2003) A Framework to Quantitatively Assess and Enhance the Seismic Resilience of Communities. *Earthquake Spectre*, 19(4), pp. 733–752.

BS. (2014) *Guidance on Organizational Resilience*. BS 65000.

Chang, S.L. et al. (2014) Establishing Disaster Resilience Indicators for Tan-sui. *Social Indicators Research*, 115, pp. 387–418.

Chang, Y. et al. (2009) Capacity Empowerment and Building: Integrated Recovery Management Framework in China. In Duncan, K. and Brebbia, C.A. (eds) *Disaster Management and Human Health Risk: Reducing Risk, Improving Outcomes*. Boston, MA: WitPress, pp. 309–322.

Chisholm, P. (n.d.) Ten Tips for Successful IT Disaster Recovery Planning. Available at: http://www.infosectoday.com/Articles/DRPlanning.htm.

Cimellaro, C.P. (2016) *Urban Resilience for Emergency Response and Recovery: Fundamental Concepts and Applications*. Bern: Springer International Publishing.

CIPedia. (n.d.) Available at: https://publicwiki-01.fraunhofer.de/CIPedia/index.php/CIPedia%C2%A9_Main_Page.

Cumming, G.S. et al. (2005) An Exploratory Framework for the Empirical Measurement of Resilience. *Ecosystems*, 8(8), pp. 975–987.

Cutter, S.L. et al. (2008a) *Community and Regional Resilience: Perspectives from Hazards, Disasters, and Emergency Management. CARRI Research Report 1*. Available at: http://www.resilientus.org/library/FINAL_CUTTER_9-25-08_1223482309.pdf

Cutter, S.L. et al. (2008b) A Place-Based Model for Understanding Community Resilience to Natural Disasters. *Global Environmental Change* 18(2008), pp. 598–606.

Cutter, S.L., Burton, C.G. and Emrich, C.T. (2010) Disaster Resilience Indicators for Benchmarking Baseline Conditions. *Journal of Homeland Security and Emergency Management*, 7(1), pp. 1–22.

Davis, I. and Alexander, D. (2016) '*Recovery from Disaster*', London: Routledge.

De Bruijne, M. and Van Eeten, M. (2007) Systems That Should Have Failed: Critical Infrastructure Protection in an Institutionally Fragmented Environment. *Journal of Contingencies and Crisis Management* 15(1), pp. 18–29.

Executive Order. (2013) *Executive Order 12636 – Improving Critical Infrastructure Cybersecurity*. Federal Register, The POTUS of America.

Flint, C.G. and Luloff, A.E. (2005) Natural Resource-Based Communities, Risk, and Disaster: An Intersection of Theories. *Society & Natural Resources*, 18(5), pp. 399–412.

Flynn, S.E. (2008) America the Resilient: Defying Terrorism and Mitigating Natural Disasters. *Foreign Affairs*, 83(2).

Ford, R., Carvalho, M., Mayron, L. and Bishop, M. (2012) Towards Metrics for Cyber Security. 21st EICAR Annual Conference Proceedings, May, pp. 151–159.

Francis, R. and Bekera, B. (2014) A Metric and Frameworks for Resilience Analysis of Engineered and Infrastructure Systems. *Reliability Engineering and System Safety* 121(2014), pp. 90–103.

142 Recovery

Galindo, G. and Batta, R. (2013) Review of Recent Developments in Or/Ms Research in Disaster Operations Management. *European Journal of Operational Research*, 230, pp. 201–211.

Gibb, F. and Buchanan, S. (2006) A Framework for Business Continuity Management. *International Journal of Information Management*, 26, pp. 128–141.

Gibson, C.A. and Tarrant, M. (2010) A 'Conceptual Models' Approach to Organisational Resilience. *The Australian Journal of Emergency Management* 25(2), pp. 6–12.

Haddow, G.D., Bullock, J.A. and Coppola, D.P. (2011) *Introduction to Emergency Management*. Fifth edn. Amsterdam: Elsevier.

Holling, C.S. (1973) Resilience and Stability of Ecological Systems. *Annual Review of Ecology and Systematics*, 4, pp. 1–23.

Hosseini, S., Barker, K. and Ramirez-Marquez, J.E. (2016) A Review of Definitions and Measures of System Resilience. *Reliability Engineering and System Safety*, 145, pp. 47–61.

HSSAI. (2009) *Concept Development: An Operational Framework for Resilience*. Washington, DC: Homeland Security Studies and Analysis Institute.

ICD-10. (n.d.) *The ICD-10 Classification of Mental and Behavioural Disorders*. Geneva: World Health Organization, pp. 120–121.

IoM. (2015) *Healthy, Resilient, and Sustainable Communities After Disasters: Strategies, Opportunities, and Planning for Recovery*. Institute of Medicine (U.S.) Committee on Post-Disaster Recovery of a Community's Public Health, Medical, and Social Services. Washington, DC: National Academies Press.

Ishiwatari, M. (2014) Institution and Governance Related Learning from the East Japan Earthquake and Tsunami. In Shaw, R. (ed.) *Disaster Risk Reduction: Methods, Approaches and Practices*. Tokyo: Springer, pp. 77–88.

ISO. (2000) Quality Management Systems – Requirements. ISO 9001.

ISO. (2004) Environmental Management Systems – Requirements with Guidance for Use. ISO 14001.

ISO. (2007) Security Management Systems for the Supply Chain. Guidelines for the Implementation of ISO 28000. 28004.

ISO. (2011) Security Management Systems for the Supply Chain – Development of Resilience in the Supply Chain. 28002.

ISO. (2014a) Security Management Systems for the Supply Chain. Guidelines for the Implementation of ISO 28000. Part 2: Guidelines for Adopting ISO 28000 for Use in Medium and Small Seaport Operations. 28004-28002.

ISO. (2014b) Security Management Systems for the Supply Chain. Guidelines for the Implementation of ISO 28000. Part 3: Additional Specific Guidelines for Adopting ISO 28000 for Use of Medium and Small Businesses (Other than Marine Ports). 28004-28003.

ISO. (2014c) Security Management Systems for the Supply Chain. Guidelines for the Implementation of ISO 28000. Part 4: Additional Specific Guidelines for Adopting ISO 28000 If Compliance with ISO 280001 is a Management Objective. 28004.

ISO/IEC. (2005) Information Technology – Security Techniques – Information Security Management Systems – Requirements. ISO/IEC 27001.

Kahan, J.H., Allen, A.C. and George, J.K. (2009). An Operational Framework for Resilience. *Journal of Homeland Security and Emergency Management*, 6(1), pp. 1–48.

Kanellopoulos, A. et al. (2014) Parental Unemployment and Post-Traumatic Stress Disorder Symptoms: A Study Through the Fog of Greek Financial Crisis. *European Psychiatry* 29(Suppl 1).

Kapucu, N. (2014) Collaborative Governance and Disaster Recovery: The National Disaster Recovery Framework (NDRF) in the U.S. In Shaw, R. (ed.) *Disaster Risk Reduction. Methods, Approaches and Practices.* Tokyo: Springer, pp. 41–59.

Klein, R.J.T., Nicholls, R.J. and Thomall, F. (2003) Resilience to Natural Hazards: How Useful Is This Concept? *Environmental Hazards* 5(2003), pp. 35–45.

Kozine, I. and Andersen, H.B. (2015) Integration of Resilience Capabilities for Critical Infrastructures into the Emergency Management Set-Up. In Podofillini, L. et al. (eds) *Safety and Reliability of Complex Engineered Systems.* London: Taylor & Francis Group, pp. 172–176.

Labaka, L., Hernantes, J. and Sarriegi, J.M. (2015) Resilience Framework for Critical Infrastructures: An Empirical Study in a Nuclear Plant. *Reliability Engineering and System Safety*, 141, pp. 92–105.

Landstedt, J. and Holmström, P. (2007) *Electric Power Systems Blackouts and the Rescue Services: The Case of Finland.* CIVPRO Working Paper, Helsinki: University of Helsinki.

LEDDRA Project. (2014) Land & Ecosystem Degradation & Diversification: Assessing the Fit of Responses. Available at: http://leddra.aegean.gr/index.htm

Lee, A.V., Vargo, J. and Seville, E. (2013) Developing a Tool to Measure and Compare Organizations' Resilience. *Natural Hazards Review* (February), pp. 29–41.

Linkov, I. et al. (2013) Measurable Resilience for Actionable Policy. *Environmental Science and Technology*, 47, pp. 10108–10110.

McAslan, A. (2010) *Community Resilience. Understanding the Concept and its Applications.* Available at: http://sustainablecommunitiessa.files.wordpress.com/2011/06/community-resilience-from-torrens-institute.pdf

McDaniels, T. et al. (2007) Empirical Framework for Characterizing Infrastructure Failure Interdependencies. *Journal of Infrastructure Systems*, 13(3), pp. 175–184.

McEntire, D.A. (2015) *Disaster Response and Recovery. Strategies and Tactics for Resilience.* Hoboken, NJ: John Wiley & Sons.

McManus, S. (2008) *Organisational Resilience in New Zealand.* University of Canterbury. Available at: http://ir.canterbury.ac.nz/bitstream/10092/1574/1/thesis_fulltext.pdf

Moteff, J.D. (2012) *Critical Infrastructure Resilience: The Evolution of Policy and Programs and Issues for Congress.* Congressional Research Service 7-5700. Available at: http://www.fas.org/sgp/crs/homesec/R42683.pdf

Nieuwenhuijs, A.H., Luiijf, H.A.M. and Klaver, M.H.A. (2008) Modeling Critical Infrastructure Dependencies. In Mauricio, P. and Shenoi, S. (eds) *IFIP International Federation for Information Processing, Volume 290, Critical Infrastructure Protection II*, Boston, MA: Springer, pp. 205–214.

Norris, F.H. et al. (2008) Community Resilience as a Metaphor, Theory, Set of Capacities, and Strategy for Disaster Readiness. *American Journal of Community Psychology*, 41, pp. 127–150.

O'Rourke, T.D. (2007) Critical Infrastructure, Interdependencies, and Resilience. *The Bridge*, 37(1).

Petit, F., Wallace, K. and Phillips, J. (2014) An Approach to Critical Infrastructure Resilience. *The CIP Report.* Center for Infrastructure Protection and Homeland Security. 12(7), pp. 17–20.

Petit, F.D. et al. (2013) *Resilience Measurement Index: An Indicator of Critical Infrastructure Resilience*. Argonne National Laboratory, U.S. Department of Energy.

Prior, T. (2015) Measuring Critical Infrastructure Resilience: Possible Indicators. *Risk and Resilience Report*, 9. ETH Zürich.

Pursiainen, C. and Gattinesi, P. (2014) *Towards Testing Critical Infrastructure Resilience*. Publications Office of the European Union, JRC Scientific and Policy Reports.

Pursiainen, C. et al. (2016) Critical Infrastructure Resilience Index. In Walls, L., Revie, M. and Bedford, T. (eds) *Risk, Reliability and Safety: Innovating Theory and Practice*. Boca Raton, FL: CRC Press, pp. 2183–2189.

Rodriguez-Llanes, J.M., Vos, F. and Guha-Sapir, D. (2013) Measuring Psychological Resilience to Disasters: Are Evidence-Based Indicators an Achievable Goal? *Environmental Health* 12(115), pp. 1–10.

Rose, A.Z. (2009) Economic Resilience to Disasters. *CARRI Research Report*, 8. CREATE Research Archive, Published Articles & Papers, pp. 7–8.

Rose, A. and Krausman, E. (2013) An Economic Framework for the Development of a Resilience Index for Business Recovery. *International Journal of Disaster Risk Reduction*, 5, pp. 73–83.

Runyan, R.C. (2006) Small Business in the Face of Crisis: Identifying Barriers to Recovery from a Natural Disaster. *Journal of Contingencies and Crisis Management*, 14(1), pp. 12–26.

Sahebjamniaa, M., Torabia, S.A. and Mansourib, S.A. (2015) Integrated Business Continuity and Disaster Recovery Planning: Towards Organizational Resilience. *European Journal of Operational Research*, 242(1), pp. 261–273.

Shaw, R. (2014) Post-Disaster Recovery: Issues and Challenges. In Shaw, R. (ed.) *Disaster Risk Reduction: Methods, Approaches and Practices*. Tokyo: Springer, pp. 1–13.

Sheffi, Y. (2005) *The Resilient Enterprise – Overcoming Vulnerability for Competitive Advantage*. Cambridge, MA: MIT Press.

Sherrieb, K., Norris, F.H. and Galea, S. (2010) Measuring Capacities for Community Resilience, *Social Indicators Research*, 99, pp. 227–247.

Sifaki-Pistolla, D. et al. (2017) Who Is Going to Rescue the Rescuers? Post-Traumatic Stress Disorder Among Rescue Workers Operating in Greece During the European Refugee Crisis. *Social Psychiatry and Psychiatric Epidemiology*.

Stephenson, A. (2011) Benchmarking the Resilience in Organisations. University of Canterbury, PhD thesis.

Sterbenz, J.P.G. et al. (2011) Modeling and Analysis of Network Resilience. *Proceedings of the IEEE COMSNETS*. Bangalore, India.

Thiruppugazh, V. (2014) Post-Disaster Reconstruction and Institutional Mechanisms for Risk Reduction: A Comparative Study of Three Disasters in India. In Shaw, R. (ed.) *Disaster Risk Reduction: Methods, Approaches and Practices*. Tokyo: Springer, pp. 17–39.

UK Cabinet Office (2013) *Emergency Response and Recovery: Non-Statutory Guidance Accompanying the Civil Contingencies Act 2004*. Civil Contingencies Secretariat. Available at: https://www.gov.uk/government/uploads/system/uploads/attachment_data/file/253488/Emergency_Response_and_Recovery_5th_edition_October_2013.pdf

UNISDR. (2009) UNISDR Terminology on Disaster Risk Reduction. United Nations International Strategy for Disaster Reduction (UNISDR), Geneva, Switzerland. Available at: http://www.unisdr.org/we/inform/terminology

Varghese, M. (2002) *Disaster Recovery*. Boston: Course Technology.

Vlacheas, P. et al. (2013) Towards End-to-end Network Resilience. *International Journal of Critical Infrastructure Protection*, 6(3–4), pp. 159–178.

Wang, C. and Blackmore, J. (2009) Resilience Concepts for Water Resource Systems. *Journal of Water Resources Planning and Management*, 135(6), pp. 528–536.

Watters, J. (2014) *Disaster Recovery, Crisis Response, and Business Continuity: A Management Desk Reference*. New York: Apress.

Wen, J.-C. et al. (2014) Typhoon Morakot and Institutional Changes in Taiwan. In Shaw, R. (ed.) *Disaster Risk Reduction: Methods, Approaches and Practices*. Tokyo: Springer, pp. 61–75.

Wilson, A.G. (2012) *Community Resilience and Environmental Transitions*. London: Routledge.

Youn, B.D., Hu, C. and Wang, P. (2011) Resilience-driven System Design of Complex Engineered Systems. *Journal of Mechanical Design*, 133(10), pp. 10108–10110.

7 Learning

While most official documents on crisis or disaster management, such as the UNISDR terminology (UNISDR, 2009), do not specifically define or even mention learning, there is a body of rather recent academic literature on learning in the crisis or lesser emergency context (for reviews, see Schiffino et al., 2017; Drupsteen and Guldenmund, 2014; Deverell, 2009; Stern, 1997). While the volume of literature is still sparse and the theoretical level somewhat underdeveloped compared to some other fields of crisis management, such as crisis decision-making, this steadily growing interest nonetheless reflects the fact that learning is increasingly understood as an essential element of crisis management proper. Some theorists even argue that the phase of post-crisis learning is more important than the actual crisis that preceded it, namely, in terms of its long-term consequences (Drennan, McConnell and Stark, 2015, pp. 192, 204). However, others do not emphasise learning as a separate phase of crisis management. This is also a legitimate point, as it is true that learning is an overarching or horizontal concept and phenomenon that is present and needed in all phases of crisis management. Indeed, one can identify debates about crisis-related learning within all phases of the crisis management cycle discussed in this book.

When one wants to discuss learning in the context of the crisis management cycle, however, this brings a certain vagueness to the concept as one of its distinct phases. Even if one were to concentrate on post-crisis learning, the different phases are difficult to keep analytically separate because learning actually materialises only in the subsequent crisis management phases. On the other hand, this is basically the situation with all the other crisis management phases discussed in this book; they can be seen as 'phases' in their own right, but they are nevertheless born from, overlap with and affect the other phases.

Another challenge is that there are different disciplinary traditions that all discuss the concept of learning in the context of crisis management. While these traditions or approaches share a common research field, or at least overlap, they proceed at different levels of abstraction and have partially different goals and problems. However, unlike different theories of crisis decision-making, for example, the boundaries between different discourses, schools of thought or theories when it comes to crisis-related learning are rather obscure, even if we were to focus on post-crisis learning specifically. As Levy (1994) noted back in the mid-1990s, post-crisis learning as a research field is a conceptual minefield, where some potentially complementary but in practice rival research agendas meet. Schiffino et al. (2017) have tried to make sense of the literature, and summarise that the power relations, the cognitive dimension of learning and its institutionalisation, as well as the social and political pressures, are all elements that contribute to post-crisis learning. Yet, the authors continue, these elements are rarely considered from an integrated viewpoint. They nevertheless identify two separate disciplinary perspectives. Researchers with a background in political science examine learning as deeply connected to accountability and blame games, that is, political struggles between competing frames in the public space on lessons to be drawn and responses to be addressed. Scholars in organisation studies, instead, have focused on the modes of learning and the barriers weighing upon organisational dynamics.

Bearing these challenges in mind, this chapter nonetheless aims at drawing a rather holistic picture, capturing in five sections the most important debates and issues in this body of literature. While the chapter starts by briefly discussing the concept in relation to all crisis management phases, the main focus is on post-crisis learning. In the first section, such conceptual and typological issues as the mechanisms, degrees and levels of post-crisis learning are discussed. The subsequent section then highlights what is possibly the most important puzzle in the literature, namely, why the post-crisis learning process sometimes fails. This is followed by a discussion on a somewhat more practical theme, namely the issue of post-crisis evaluations or formal investigations, which are usually conducted after major crises. The chapter concludes by taking up the blame games and political wrangling that often follow major crises, as a form of distorted post-crisis learning rituals.

7.1 Learning as a horizontal crisis management challenge

As mentioned above, learning is, in one way or another, related to all phases of crisis management. First, it is obvious that learning from previous crises is an essential element of risk assessment (Moynihan, 2008, 2009). Properly conducted risk assessment always relies on or utilises existing historical data to identify the risks and estimate their probabilities. If this historical data has been left unrecorded, it is a learning failure that negatively affects risk assessment and future crisis management as a consequence. A past crisis may also have introduced new risks or new dimensions of risks that were previously omitted and trigger learning processes that take the past experience into account when assessing risks in the future. Basically, the ISO 31000 risk management standard takes this explicitly into account, expecting that the risk assessment phase is continuously monitored and reviewed in light of new data.

By this very logic, learning is also naturally present when introducing preventive and mitigating barriers or controls to the risks that were identified in previous crises, such as vulnerabilities and interdependencies that did not occur to the risk assessment or crisis management team. Rerup (2009), among others, however, emphasises the limitations of the standard reinforcement learning processes of simple trial and error. The issue boils down to avoiding the error in the first place. Prevention, by definition, should comprise learning proactively from issue-specific weak cues based on previous experience, thereby avoiding the crisis. In this conception, learning can indeed be seen as an early warning-based prevention strategy rather than a post-crisis phase, although the difference is largely semantic. Veil (2011), focusing on business crises, also emphasises the capacity of learning to read the weak signals of emerging crises, and this capacity or capability then becomes a preventive tool for avoiding future crises. Similarly, as suggested by Deverell (2015, p. 200; cf. Westrum, 1996, 2004), organisational safety culture, for instance, can be seen in terms of a set of values and practices that may enable preventive action by learning from previous crises or near-misses before any new crisis emerges.

Learning can also be understood as an important element of preparedness. Training and exercises, as discussed in the respective chapter in this book, are central parts of preparedness. It is obvious by definition that the motivation to organise different types of training, simulations, drills and exercises is to learn from them. Previous crises are often used as a basis for scenarios or lessons learned to equip trainees with the requisite skills to put into practice when facing a crisis. Learning from training and exercises is, however, not always automatic and rarely optimal. Consequently, there is a rather lively ongoing discussion about the kind of training, tools and conditions that produce useful and tangible learning outcomes for different aspects of crisis management (e.g. Borella and Eriksson, 2013; Fabbri and Chung, 2009; Borodzicz and Van Haperen, 2002; Carrel, 2000). On the other hand, post-crisis learning as such can also be understood as a preparedness strategy. Carmeli and Schauerbroeck (2008), among many others, have identified a generic positive association between learning from past failures and crisis preparedness.

Learning also takes place in the response phase. Moynihan (2008, 2009) discusses this in terms of intra-crisis (during the crisis) learning, compared to inter-crisis (after the event) learning. The argument is that intra-crisis learning is more difficult, due to the very characteristics of a crisis. During the crisis situation, learning is also constrained by high stakes, uncertainty and time pressure. Lessons have to be drawn under conditions of limited information and with little sense of how probable available alternatives are likely to be. Urgency in particular can lead to ill-considered lessons, as decision-makers do not have the same luxury of carefully calculated conclusions as post-crisis evaluators. Müller-Seitz and Macpherson (2014) further argue that learning during the crisis response phase is also characterised by the contested and uncertain nature of what is happening as different stakeholders tend to react from the perspective of their own specific epistemic commitments. This means that they are likely to translate their experiences into organisational and institutional responses that make sense from within their own paradigms, but not necessarily from within other viewpoints. Kamkhaji and Radaelli (2017,

p. 722) adopt the concept of intra-crisis learning, emphasising that more research is needed to understand the phenomenon of learning under uncertainty and time pressure. However, referring to Birkland (2009), they admit that this kind of during-the-crisis learning is often about finding good interim solutions to unanticipated problems that cannot be resolved through standard operating procedures; in other words, such behaviour can also be seen merely as improvisation.

Finally, learning can be understood as part of recovery, as already discussed in the previous chapter. The UK approach to civil protection (UK Cabinet Office, 2013) deals with learning mostly in terms of recovery efforts. The focus is on failures, using the lessons learned from actual and previous crises in this phase of the crisis management cycle. As Shaw (2014, p. 2) puts it:

> Quite often, post-disaster recovery leads to rebuilding risk; recovery efforts are not informed by lessons learnt and experiences from previous disasters; recovery-needs assessment has not been demand-driven; stakeholder consultative processes are weak; institutions set up to manage recovery have not led to sustained national and local capacities for disaster reduction.

In other words, one should learn from previous crises, examining their situational and systemic causes, and not rebuild the pre-existing conditions for a crisis and thereby prepare the ground for future crises.

However, while it is somewhat unclear where crisis-related learning starts and where it ends, most of the literature discusses learning as a phase following the acute phases of response and recovery. Crises, and especially failures to prevent or deal with them, reveal weak points in a system or organisation, which must be corrected in order to prevent the same failures from recurring (Deverell, 2015). In this sense, failures are paradoxically welcomed. The normative lesson from the learning-from-crisis literature is precisely the idea that smart people have acquired wisdom through failure, whereas one acquires very little wisdom from success (e.g. Hällgren and Wilson, 2011).

7.2 The mechanisms, degrees and levels of learning

Part of the post-crisis learning literature is rather theoretical in spirit, discussing such issues as whether this type of learning differs from 'normal' learning, what kind of learning it is, and even such a basic question as who learns when we speak about post-crisis learning. This section highlights some main points in this body of literature.

Is post-crisis learning different from normal learning?

In very generic terms, learning is about knowledge. Sommer (2015), basing his account on a review of a broad body of literature, distinguishes between conceptual knowledge, procedural knowledge and dispositions. Conceptual knowledge is about 'knowing that'. It comprises information, facts, assertions, propositions and concepts. In the context of a crisis, it could refer to a declarative understanding of what to look for when assessing a crisis situation, or why a certain kind of behaviour is appropriate. This kind of knowledge is explicit in nature, and is not acquired through personal experience but through the communication of generalised knowledge based on someone else's experience. Procedural knowledge is about 'knowing how'. It comprises the techniques, skills and abilities to attain certain goals which, in our case, pertain to the crisis management field. In a crisis, the goals may be known but, due to the inherent nature of the crisis situation, also unknown. This kind of knowledge about how to do things can be regarded as practical knowledge, and is often implicit. Dispositions, in turn, refer to a deeper level of knowledge underpinning conceptual and procedural knowledge. They comprise such issues as interests, values, attitudes, emotions and personal motivation. In a crisis situation, dispositions comprise individuals' tendencies to put their capacities into action. Some of the dispositions are purely individual, whereas others are related to one's profession or specific workplace.

While some concepts used in the literature of post-crisis learning derive from the huge body of more generic learning literature, the latter is often difficult to apply to post-crisis learning. Deverell (2015) summarises the views of most post-crisis learning experts by stating that learning from crises is different from learning in general. It is triggered by a certain unexpected or harmful event, and is not only the normal, cumulative long-term adaptation to changing circumstances or new information. It is also more difficult, as even post-crisis learning – and not only during-the-crisis learning – is often characterised by the same factors as the crisis itself, such as a threat to important values, uncertainty, and time pressure. In larger-scale societal crises, or in business crises, the media, stakeholders, shareholders, political rivals, and so forth, carefully monitor not only how the acute crisis was dealt with, but also the lessons that were learned from it. Quick answers and solutions are demanded, scapegoats are sought and the political, bureaucratic, financial or reputational survival of those considered to be responsible is at stake.

Post-crisis learning, should this take place, is important because one of its main characteristics is that it entails change in personal and/or organisational behaviour. This post-crisis change is different in nature compared to change during normal times. Kamkhaji and Radaelli (2017) make a distinction between more normal-time adaptive learning and crisis-related contingent learning. While adaptive learning considers the situation from a more cause-effect perspective, in and after a crisis situation a more psychological, even unconscious learning process can take place, where the unexpected elements and uncertainty prompt the actor to change their behaviour and sometimes even their mindset. Otherwise resilient belief systems may not pose a major hindrance to learning and change, because surprise may trump prior beliefs via self-reinforcing mechanisms of association between new stimuli and outcomes. Schiffino et al. (2017) similarly argue that a crisis triggers a new game, a new set of problems in the face of which actors find themselves mutually interdependent. In this new game, actors strategically engage in interactions to determine the lessons that are to be drawn, the decisions that are to be taken and how change should be effected. In short, post-crisis learning results in new collective rules of the game that are socially produced.

Carmeli and Schauerbroeck (2008) add that learning effectively means that previously learned behaviours need to be unlearned. This, they argue, requires building a culture of learning from failures, which provides new responses and discards old ones, seeking not only to correct the immediate failure so as to ensure the continuity of an operation or the provision of a service, but also to address the root causes of the problem.

The general notion, then, is that post-crisis learning may lead to a major change more readily than normal-situation adaptive learning. It is often the case that only a crisis can break the so-called path dependence of a person, organisation or system that has become too used to self-reinforcing its routines, behaviour and values, and whose past decisions may restrict a change in normal conditions. On the other hand, a crisis can also be used to deliberately enhance a change that would not have been possible or even imaginable otherwise.

Whether learning actually takes place, and through which mechanisms, is then another puzzle. Boin et al. (2005, pp. 117, 188) discuss post-crisis learning capacity and differentiate between three types of learning: experience-based, explanation-based and skills-based. First, direct lessons from a crisis require some kind of individual or organisational system of memory that records and recollects the experiences; that is, what has happened and how the actors responded. These kinds of memories are then transferred to lessons. Some organisational theorists regard this type of learning from direct experience as the only way to produce real results: "Only when direct experience serves as the basis for learning can educative experiences be realized" (Kayes, 2015, p. 150). Second, explanation-based learning, according to Boin et al. (2005), relies on experts who are capable of explaining the cause-and-effect relationship related to the crisis. While in many cases this is relatively easy, in other cases it might demand months or years of investigations. Third, skills-based learning is triggered by a crisis when a new type of phenomenon is behind the crisis. This might be, for instance, a new type of pandemic or computer virus, which then leads to studies and learning processes to build up better capability and capacity to face similar crises in the future.

Sommer and Lussand (in Sommer, 2015, pp. 253–295), in their case study on Norwegian police officers' learning, provide much more concrete and practice-oriented categories or mechanisms for learning. Their case study provides an example of how difficult it is, on the one hand, to prepare for unexpected crises, and to learn from them, on the other. Their research involved comparing the learning system before and after the major one-man terrorist attack launched against the Government Complex and subsequently against the Labour Party youth camp on Utøya Island in 2011. They go through the learning mechanisms prior to the attack: initial training and education; regular training; knowledge storage and sharing; informal storytelling and discussions; debriefing sessions; and formal post-crisis evaluations. Despite the fact that this system seemed to work well during normal police work, when faced with a major crisis, it proved inadequate in preparing the police force properly; official post-crisis investigation reports criticised the activities and preparations carried out by the police rather heavily. The problem was that most of their training had focused on normal police work and minor emergencies. The authors found that more attention was paid to learning after the crisis, the aim being to enforce and systematise it in all phases of crisis management. However, in 2015, four years after the crisis, the authors concluded that "this work related to learning is… in its initial phase, and therefore it is too early to conclude on how it will affect emergency response capabilities in the police services" (Sommer, 2015, p. 289).

Fast and slow learning

Learning can also be distinguished according to different degrees, in line with theories originating from behavioural science and subsequently expanding into other fields. For instance, in international relations or foreign policy analysis, fields that have explored the concept for decades (e.g. Levy, 1994), one talks about 'fast' or 'simple' learning, on the one hand, and 'slow' or 'complex' learning, on the other. This distinction is related to the degree of change that learning will cause. Fast or simple learning is nothing more than an immediate feedback reaction to the crisis and its management, and is reflected only in a new choice of means. The goals remain the same, but one makes some tactical or technological revisions, enhancing efficiency, in order to achieve these goals. Slow or complex learning, in turn, takes place more at the level of beliefs and values, and can lead to a redefinition of the problem or a revision of the goals. Both what is done and the way it is done will be changed.

In more specific crisis management literature, Birkland (2006) uses the terms instrumental, societal and political policy learning. This is basically the same as what Drennan, McConnell and Stark (2015, p. 204) refer to as finetuning, policy reform and a paradigm shift. Fine-tuning consists only of small-scale tweaks to pre-existing policies; policy reform is a process of slow adaptation that involves new policy principles and new institutional values; and a paradigm shift takes place when the basic objectives and goals are changed.

The most commonly used concepts in the crisis management literature, however, referring to the same issue, were coined by Argyris and Schön (1978, p. 2), who use the terms single-loop learning and double-loop learning, representing fast and slow learning, respectively. The former concerns an organisation detecting and correcting an error, while the latter takes place when the detection and correction of the error involve modifying the organisation's underlying norms, policies and objectives. The theorists claim that a major shortcoming in organisations is that they often attend only to single-loop learning, neglecting to look into the root causes of the errors or identify the new behaviours needed to prevent reoccurrences. Bearing this stance in mind, Carmeli and Schauerbroeck (2008) argue that many crises emerge and evolve because system failures interact to generate a crisis situation, and that these failures could be addressed far more easily if double-loop learning were employed. Johansson (2016) adds triple-loop learning to this equation, which is basically the same as Drennan, McConnell and Stark's (2015) notion of a paradigm shift.

As noted above, Moynihan (2008, 2009) complicates this picture with his division between intra-crisis (response) and inter-crisis (post-crisis) learning. While single-loop learning in an intra-crisis context might mean a change of tactics, double-loop learning would imply a change of strategy that might result from a significant mismatch between contingency plans and reality. In the inter-crisis context, single-loop learning could mean modifying systems and policies across time, such as standard operation procedures, whereas double-loop learning could lead to institutional changes.

Fast and slow learning do not have to be mutually exclusive, however. Roux-Dufort and Metais (1999), who focus on the French nuclear industry, provide an example of how a company can learn not from their own crisis management but from other experiences in the field. The objects of learning are the lessons learned from Three Mile Island in 1979 and Chernobyl in 1986, which served to improve and enrich the core competences in risk and crisis management. The authors distinguish between three chronologically consecutive phases in the learning process, namely, the technical phase, the human phase and the cultural phase, representing the different degrees of learning discussed above to some extent. The starting point at the time was underpinned by the belief that the management of a plant essentially relied on technical considerations. This belief was severely shaken by the above-mentioned two major events, which shifted attention away from technology to the management of that technology. The focus was now directed towards the significance of human error and the necessity to constantly re-evaluate the ergonomics of control rooms. The erstwhile almost blind confidence in technology that had undermined the role of human beings in complex technologies was duly channelled into detecting warning signals and anomalies, providing more scope for human evaluation and early intervention. After a while, the whole culture started to change, and control was no longer determined solely through the hierarchy and rigid procedures, but equally by the expediency of shared values, and the implementation of a true safety culture.

Who learns?

The question then arises about the level of learning, that is, who learns when we speak about post-crisis learning? The literature generally shares the fundamental argument that organisations are not regarded as capable of remembering, thinking or learning by themselves; it happens at the individual level. Learning by an organisation is therefore often regarded as a multi-stage process. Feedback from the environment makes an individual or several individuals learn, which in turn prompts these individuals to try to change the activity of the respective organisation, which may (or may not) lead to a change in the organisation's behaviour. This in turn will evoke a new wave of feedback.

Deverell (2015) argues that an individual's learning is a necessary but not sufficient precondition for a change of practices and policies through organisational learning. Learning from crises calls for individuals who are conscious and strategic in their approach to learning from the past crisis, and who have to be motivated and able to drive the required reforms throughout the organisation. Because this individual learning takes place within and for the organisation, it can, in some cases, be transferred into collective learning processes even before it turns into organisational learning. At the same time, the existing organisational culture and practices, both formal and informal, set limitations on this learning in terms of obstacles and opportunities for change.

One often speaks about a three-phase process: lessons identified, lessons learned and lessons institutionalised (e.g. Milton, 2010). It could be argued that the first phase of this process takes place at the individual level, and that the second possibly already includes the organisational level, whereas an organisation has truly learned only when the change has been institutionalised in the organisation's culture and behavioural practices. Deverell (2009), in turn, makes a distinction between the concepts of lessons distilled and lessons implemented. In the former, lessons are more like a cognitive activity, whereas the latter is a behavioural activity. A crisis-induced lesson is then distilled when new information or knowledge based on the crisis experience is declared in statements by organisational members, which may be for real or rhetorical reasons, however. The lesson is considered implemented when it leads to a systematic alteration of behaviour.

In addition to individual and organisational learning, Moynihan (2009) introduces the concept of network learning, which is particularly important in the context of a crisis. With a public sector crisis in mind, such as a large-scale societal, man-made, natural or technological crisis, he rightly concludes that any major crisis depends on more than a single organisation. Instead, these types of crises typically involve a network of multiple government jurisdictions, as well as private and nonprofit actors. Network learning is basically the same as what Johansson (2016) refers to as different societal levels being involved in learning processes, such as the EU, the respective national government and parliament, companies, corporate management and employees, and so on.

While Moynihan (2009) does not provide any clear theory about this type of learning, he nonetheless notes that it creates an additional complication for the study of learning concepts, especially in the absence of a developed body of literature on network learning. Moynihan remarks that as networks are more heterogeneous than hierarchies, each actor brings their own cultural and institutional background to bear on the learning process. This, in turn, has the

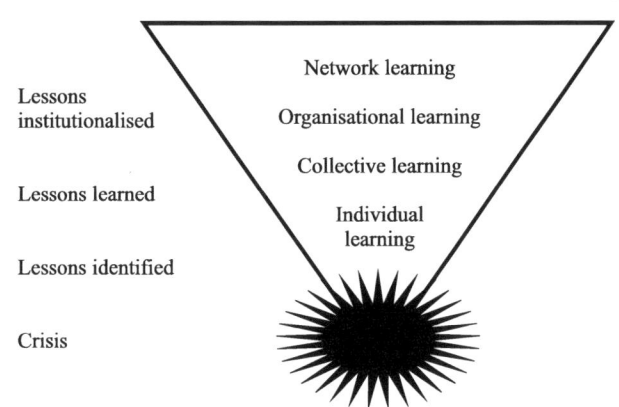

Figure 7.1 Levels of post-crisis learning

potential to create competing interpretations and conflict, and does not necessarily enhance learning.

The main levels discussed in the crisis management literature are illustrated in Figure 7.1. While the literature about the different levels of post-crisis learning can be considered an under-researched area, especially concerning the mechanisms and conditions through which individual lesson identification is transposed into institutionalised lessons, this typology nonetheless creates a basis for further research and hypotheses.

7.3 Failure to learn

Common sense would suggest that one should learn from previous crises in order to manage future crises better. However, as the saying goes, common sense is not very common. We can find numerous narratives and analyses describing the failure to learn from crisis outcomes, or at least expressing criticism towards the learning process. In analytical terms, one can distinguish at least two different types of failure in post-crisis learning. First, one does not learn anything from past crises, or learns only superficial things that do not lead to any changes or reforms. Second, one learns, but the wrong lessons.

Why do some not learn?

Indeed, post-crisis learning is often seen as a neglected phase of crisis management:

> Many companies spend thousands of people-hours on planning and millions of dollars on implementing but very little time reflecting on what they have done. They don't approach learning [from crises] in a systematic way. Consequently, they lose much of the value that comes with experience.
> (Harvard Business School, 2004, p. 114)

156 *Learning*

Elliot (2009), focusing on two serious failures in child healthcare issues, argues that the organisation in question basically repeated the errors experienced in the first case and had not learned anything at all, despite a thorough public evaluation. He concludes that the policy (lessons learned) emerging from the processes surrounding the public inquiry in many societal crises may be successful, but there is a gap between policy and practice (lessons institutionalised) that constitutes a bottleneck, and which has received insufficient attention to date.

While a crisis may be a catalyst for learning as Deverell (2009) argues, or open a window of opportunity for policy change as Hansén (2015) and Boin et al. (2005) propose, it is not automatic. From an analytical perspective, the issue revolves around the question of why the learning-from-crises process fails in some cases, but succeeds in others, and through which mechanisms. In fact, this is a research question that guided those theorists that raised learning from crises as a specific topic in the crisis management literature (e.g. Lagadec, 1997). Numerous factors are suggested in the literature to explain post-crisis learning failures. Reviewing this literature gives an impression of a rather fragmented research field. Some of the sources or factors contributing to failures in learning from a crisis are summarised in Table 7.1 and briefly discussed below. The list illustrates a research field in which many gaps may well exist vis-à-vis explanatory factors, and which does not necessarily form an integrated theoretical basis. While most of the factors are not directly contradictory but complementary, they are often case-specific.

The most obvious reason, proposed by most who write about the failures of post-crisis learning, is organisational inertia. Sometimes the dominant beliefs prevail and not even a fine-tuning of existing policies can be identified. As early as the 1970s, Turner (1976), in his classic study on the failure of foresight, concluded that rigidities in institutional beliefs contribute to failures that lead to disasters or crises. His study directly concerns learning, but the phases of a failure start from initial culturally accepted beliefs about

Table 7.1 Factors contributing to learning failures

Factors (examples)	*Why (examples)*
Organisational inertia	Policy resistance in complex systems, lack of organisational trust
'Elite escape'	The broader policy context blocks changes
Characteristics of crisis	E.g. external crises more likely to lead to changes
Characteristics of decision-makers	Reformists vs. conservatives; greed or other negative characteristics of top leaders
Power relations	Minority vs. majority

the world and its hazards, with associated precautionary norms set out in laws, codes of practice, mores and folkways. Sooner or later, the accumulation of an unnoticed set of events occurs, which is at odds with the accepted beliefs about hazards and the norms for their avoidance. If these latent characteristics of a system make a crisis possible or even inevitable, it should lead to – in the case of positive learning – full cultural readjustment, where beliefs and precautionary norms are adjusted accordingly in a double-loop fashion.

However, this kind of organisational learning is difficult. Sterman (2006), for instance, focusing on policies to promote public health and welfare, argues that while evidence-based learning should prevent policy resistance towards the required changes, learning in complex systems is weak and slow. Learning often fails even when strong evidence is available due to erroneous but self-confirming inferences, allowing harmful beliefs and behaviours to persist and undermining the implementation of beneficial policies.

Some other explanations have also been proposed. If the question concerns a major crisis with political dimensions and the policy-makers obviously draw no lessons from it, Hansén (2008) calls this failure to learn 'elite escape'. He explains this by arguing that crises do not take place in isolation, but have to be seen in the context of other events. The context, for instance, of approaching elections may affect whether or not it is beneficial to draw drastic lessons from a crisis. Boin et al. (2005, pp. 126–129) are rather relative in their account. They argue that at least three factors – decision-makers, crises and the respective power relations – vary in their character. Depending mainly on these issues, decision-makers may be open to changes or, alternatively, opt to refrain from carrying out any reforms. The authors divide decision-makers into reformists and conservatives, the former being proactive in proposing changes and alternatives compared to the pre-crisis situation, the latter defending and maintaining the existing institutional arrangements and showing readiness only for superficial changes. Of these two types of decision-makers, only the reformists are capable of slow or double-loop learning, and ready for profound changes in defining the problems and goals. The authors further develop four hypotheses about conditions where decision-makers are more likely to adopt a reformist double-loop approach: when the crisis is a result of shifts in the external (e.g. war, climate change) rather than the internal environment; when they are convinced that long-term problems can be solved by adopting a proactive role towards changes; when they perceive that they have a sufficient majority backing their reformist policy line; and when they feel that they can utilise the opportunity for ad hoc centralisation of power and authority due to the crisis. Should one or more of these conditions be absent, in all probability no post-crisis learning, or at least more profound learning processes and the respective changes, are likely to occur.

In more practical or concrete terms, certain case studies may shed light on the mechanisms of learning failures. In Hansén's (2015) case study, the crisis concerned the 1975 West German embassy seizure in Stockholm by terrorists.

In the aftermath, the police demanded the establishment of a specific anti-terrorist force. In other words, the police were ready to draw lessons and demanded an institutional change to be better prepared in the future. However, the politicians, the ultimate decision-makers in this case, preferred the status quo, illustrating Hansén's 'elite escape' argument. Only in the aftermath of the 1986 assassination of Prime Minister Olof Palme was this kind of special force finally established, even if the murder was apparently not a terrorist attack.

Stead and Smallman (1999), for their part, have studied business crises, especially the question of why one so often finds examples of post-crisis learning failures insomuch as companies are prone to repeating the same errors all over again. They conclude that the organisational culture can be singled out as the main contributing factor in these failures. Sommer (2015) arrives at the same conclusion in relation to learning in emergency response work, which concerns a low degree of systematic knowledge accumulation and experience-sharing in emergency management organisations, reflected in the absence of post-response evaluation practice and only rarely implemented systems of feedback on the performance of emergency personnel.

In a rather concrete evaluation focused in a similar vein on civil protection or emergency management, the Swiss CSS (2016) also emphasises the role of organisational culture. While any crisis or emergency results in positive or negative outcomes, the organisation must still be able to interpret these experiences, assisted by the creation of a collective memory that is readily accessible, and which in turn fosters a culture of learning. In order to learn, an organisation must be flexible, which encourages an open and adaptable organisational culture. The authors propose that learning itself can be seen as either outcomes or processes. If considered an outcome, the lesson automatically exists as the end product in a process of evaluation or reporting. But only if a lesson is identified as an element in a continuous process can it become a driver of change, where the change is visible at the endpoint of the process. They further highlight that knowing how to learn is, strategically, at least as important as knowing what to learn, the former being, in a way, a deeper or more structural strategic goal.

If so-called elite escape, proposed as one explanation for learning failures, can be understood as rather tactical opportunism and contextualised behaviour, organisational culture is clearly more related to slow or double-loop learning. It is slow, as Stead and Smallman (1999) note, because the change in organisational culture cannot be implemented overnight. Even if such a learning process were triggered by a crisis, the change must be carried out in stages, with the commitment of the top management of the respective organisation. They also claim that such a change must be flexible and open to later amendments. Such issues as groupthink, discussed under the response phase of crisis management, are difficult to overcome as both situational mindsets and more structural institutions are involved.

Drupsteen and Guldenmund (2014), focusing on incidents and near-misses rather than major crises, discuss the conditions for learning by identifying both the main facilitators, as well as barriers to learning. First, they pay attention to organisational trust. In respect of the literature and experiences of safety culture, they argue that if the aim is to learn from an event, no blame should be apportioned to the actors involved; people need to feel comfortable about reporting what happened. A climate of openness will make people more willing to report and discuss errors, and in so doing to learn more about the system in the process. An absence of trust, on the other hand, may lead to faulty reporting, complete lack of reporting, secrecy and less transparency, or fear of blame or social sanctions, which automatically fosters reluctance to report and therefore limits the information available to draw lessons from. Second, Drupsteen and Guldenmund emphasise that the learning process is also influenced by the incident itself and the importance that is attributed to this incident by the organisation or its stakeholders. Incidents with severe consequences easily garner media coverage and create external pressure, which motivates individuals to draw lessons and make operational and cognitive adjustments. Indeed, the authors claim that only if errors result in relatively severe consequences is a more profound need for learning perceived. Although learning is possible after many events, more attention is paid to incidents with a major impact. Finally, the authors note that individuals also influence the learning processes. Their attention is not on the decision-makers, but rather on those who can provide an account of what happened. In investigations of events, interviews are the main source of information. However, human memory is fallible and interviews are biased by hindsight. What is more, people may feel reluctant to share information about the course of events leading up to an incident.

Cheng and Seeger's (2012) case study concerns the biggest company failure in Australian business history, which led to losses of several billion dollars. The crisis was due to the company's mismanagement of its core business services, namely, insurance policies. Several top-level directors were consequently prosecuted and sentenced for their unethical and illegal behaviour, which then created an organisational crisis out of the original financial crisis. In short, they tried to cover up and lie about their wrongdoings, but could not escape widespread public criticism. Cheng and Seeger (2012) draw post factum lessons from the crisis. First, the warning signals were totally ignored because the board did not receive correct and sufficiently detailed information. Second, the company failed in their external communication. They were led in accordance with the philosophy that corporate information is private and not subject to public disclosure. However, as a listed company, their main sin was the failure to publicly disclose the company's deteriorating position, as it might have impacted their share price. As openness and transparency were lacking, and essential financial information was withheld, the stakeholders and the general public soon adopted a more hostile position, towards which the top management continued to react defensively. Third, the company never voluntarily accepted responsibility for its decline and eventual liquidation. The accused denied any wrongdoing, and the CEO soon earned the title of

Australia's most-hated man because of his lack of business ethics. This example, according to the authors, shows that, in the business world, the denial of obvious wrongdoing, shifting the blame onto others and the refusal to take responsibility for one's actions are not ethically sound strategies if one is aiming to avoid reputational damage, which may be irreversible. Thus, corporate executives need to identify what they regard as the basic moral underpinning of their system of values in order to safeguard their reputation and the company's credibility.

In any case, if we accept the equation of lessons identified leading to lessons learned, pointing to lessons institutionalised, we can duly conclude that failure to learn may occur in any of these phases, individually or in combination.

The wrong lessons?

However, when post-crisis learning takes place, it may often turn out to be superficial or, worse, the wrong lessons may be learned. The September 11, 2001 terrorist attacks in the United States have served as an example of both. Birkland (2006) investigated the learning outcomes after this event. He identified dozens of swift new regulations and institutional changes that were introduced as a consequence of the attack, but concluded that while they represented a policy change, there was little evidence of policy innovation; most of the new ideas were not new at all, but simply laws and regulations containing all the previously proposed measures. Furthermore, he concluded that when it came to natural disasters, policy changes focused on recovery problems only, and did not deal with preparedness and other phases of the crisis management cycle. The argument that, due to this event, the United States leadership overreacted by focusing all of its attention on the war on terrorism, and forgot that there were other crises to attend to as well, is almost a commonplace in journalism; this consequently served to explain the partially failed and much criticised preparedness and response to Hurricane Katrina in 2004. One can find some evidence to support the claim that the latter event did indeed result in a kind of double-loop learning, leading to a shift from an overly dominant counterterrorism approach towards a more balanced all-hazard approach. Moteff (2012), in his policy paper to the United States Congress, states that in 2006 the Critical Infrastructure Task Force of the Homeland Security Advisory Council initiated a public policy debate in which the main theme was that the government's critical infrastructure policies were overly focused on protecting assets against terrorist attacks, and inadequately focused on improving the resilience of assets against a variety of threats. In 2008, this in turn prompted the House Committee on Homeland Security to hold a series of hearings addressing this paradigm shift.

Hansén (2009), for his part, has paid critical attention to the role of buzzwords in enhancing learning in crisis management. His case study on the role of 'shared situation awareness' in Swedish crisis management was drawn from the military field and seemed to provide the key to solving the cooperation, coordination and communication problems that were perceived to exist in previous crises. The concept rapidly became a new cross-sectoral doctrine and reached all levels of governance. However, this learning outcome brought a new set of problems with it. The focus on shared situation awareness obscured the view from alternative interpretations of crisis situations. When a crisis emerged, alternative interpretations were blocked in pursuit of a shared interpretation. The political will resulted in institutional arrangements in which the concept of shared situation awareness became crucial, leading to enhancing the traditional Swedish administrative characteristics, namely, consensus and corporatism, which limited crisis management opportunities.

Crisis managers are not the only ones who might learn the wrong lessons. Bennett, Chiang and Malani (2011) focus on 'social learning', particularly during a crisis, namely, on how the public at large behaves in a crisis situation and on which information basis. While their case concerned how and what people learned during the SARS epidemic in Taiwan in 2003, their problematique is relevant to any major crisis, from nuclear accidents to stock market crashes. The authors note that people learn both from public and private sources, and that the latter in particular may pose a problem; what people learn from their peers may strongly influence the duration and severity of a crisis in that their perception of risk may deviate from reality in either a positive or a negative direction. Therefore, controlling public learning through effective crisis communication is essential for a crisis manager.

These examples suggest that while learning from crises is important, one should not overlook the possibility of learning the wrong lessons. One of the most common examples of this is expecting the next crisis to be of the same nature as the most recent one, and hence putting all one's efforts into combating a similar crisis. However, a crisis is defined by its very unexpectedness and uniqueness. Indeed, CSS (2016) point out as one of the main conclusions of their case studies that crises by definition are often about never-before-experienced hazardous events, in which case past experiences should never become the sole source of learning.

7.4 Post-crisis evaluation

An important feature of an effective organisation is the presence of an institutional system for learning from past experiences. In many cases, public and also private actors are required by law to conduct evaluations after crises or near-crises. In the case of crises, or even smaller incidents that have larger societal salience, the media are the first to demand a public or impartial inquiry. In any case, from a normative point of view, post-crisis evaluations should be an organisational routine and the evaluation results should be taken into account in developing further activities (Johansson, 2016).

Evaluations usually consider the causes and outcomes of a crisis, and the circumstances in which it occurred, as well as the related crisis management phases – for the most part, preparedness and response. Indeed, the biggest lesson to be learned from many post-crisis interviews is that one needs enhanced preparedness in terms of experience, either in the form of drills or real experience (e.g. Dawes, Cresswell and Cahan, 2004, p. 63).

But is this kind of learning outcome a valid conclusion, or is it just an answer that is self-evidently given when this issue is broached in interviews in a post-crisis situation? One of the problems is that there is no standard or shared guidelines on how to conduct post-crisis evaluations. In the academic literature, one can discern some efforts in this direction, however. McConnell (2011) is one of those who has tried to benchmark post-crisis evaluations. He divides the evaluation into processes, decisions and politics, defining criteria for success and failure under each of these categories. In practice, however, evaluations are much more pragmatic. Grunnan and Maal (2015), focusing on natural disasters and combining the identified lessons from several individual evaluations, categorise the lessons learned into eight problem areas: knowledge, interoperability, prevention, communication, risk assessment, management, training and logistics. Under each of these areas, they then identify several lessons that could generally be seen as best or good practices.

A second problem concerns the subjective nature of evaluation, at least in many cases, which again is largely due to the lack of standards. Elliott (2009), referring to several empirical studies, argues that public evaluations or inquiries are almost by definition not necessarily impartial. They are political in nature, and an underlying purpose might sometimes even be to protect vested interests. McConnell (2011) has listed other challenges that a post-crisis evaluation often entails: differences in perceptions; the lack of agreed-upon benchmarks; the winners and losers of many decisions; unclear boundaries of the evaluation; differences in results depending on whether one evaluates short-, middle- or long-term outcomes; often unclear goals in the midst of a crisis turmoil; and the impossibility of properly evaluating the alternatives that were not chosen. Similarly, Drennan, McConnell and Stark (2015, pp. 193–198, 208) note several problems posed by preparing a proper evaluation or account of what actually happened. This subjectivity already starts when defining the basic goal of the evaluation. Some evaluations are post factum accounts of the causes of the crisis and the efficiency of crisis management, while others focus on political gains.

If the latter is the case, it leads to power struggles, pre-existing views, reluctant witnesses, political interference, lack of material and non-material resources for investigation, and even actions leading to the cover-up of sensitive issues. While there are naturally learning processes in clear political crises due to domestic conditions (e.g. Grosjean, Ricka and Senik, 2013), or foreign policy experiences (e.g. Levy, 1994), even non-political crises can become the subject of political struggles. Broekema (2016), among others (e.g. Boin et al., 2005; Boin, McConnell and 't Hart, 2008), has used the term politicisation of a crisis, which, according to him, interferes in the learning processes deriving from crises either positively or negatively. Politicisation takes place when a crisis suddenly changes into crisis politics in which major interests are at stake. Politicisation, according to Broekema, stems from disagreement over interpretations of, in particular, the course of events, the underlying causes and effects, questions of responsibility and accountability, and the lessons that should be drawn.

A third problem is that the lessons often remain at the statement level only, without implementation in practice. Boin et al. (2005, pp. 84–86) go as far as to call most post-crisis investigations 'rituals' that are designed to convince the general public that decision-makers are taking the crisis seriously. Birkland's (2009) highly critical account also suggests that in many cases – especially if political goals are involved – learning processes are not taken seriously; rather, they are just a mere show for the public. Once the investigation is complete, the actors tend to reject the lessons on the grounds of cost, feasibility, or for some other reason, or will simply ignore them altogether. Moreover, such processes simply result in reports that fail to address the real problems revealed by the crisis. In a very similar spirit, Johansson (2016) argues that it is often also a question of a lack of communication plans about the lessons learned. The end result of an evaluation is usually a report, but the lessons are not communicated properly to the respective stakeholders, and they remain hollow documents. As Elliott (2009) concludes, this state of affairs leads to the conclusion that more attention would be needed, focusing on both barriers to, and facilitators of, learning from post-crisis evaluations, with the aim of translating new understandings into revised norms and behaviours within organisations.

One way of distributing the lessons learned, if not completely institutionalising them, is to organise workshops with the main stakeholders. This was done, for instance, by the United Nations in cooperation with the most affected countries after the East Asian tsunami in 2004 (e.g. Government of Indonesia and United Nations, 2005; Government of Sri Lanka and United Nations, 2005; Ferreira Pedroso et al., 2015). In the case of Sri Lanka, for example, participants joined four working groups handling the institutional and legal framework; the response mechanism and stand-by arrangements; coordination; and early warning and awareness-raising. Under each working group, several lessons were identified, both in terms of success and failures, with the aim of correcting the structural conditions that made the failures possible, or at least identifying the crisis management bottlenecks. However, there seems to be an obvious research gap when it comes to acquiring more knowledge about the degree to which, the conditions in which and the mechanisms through which the lessons learned turn into institutionalised lessons that will influence institutions and their behaviours in a positive way in their crisis management capacity and capability.

7.5 Blame games and management by crisis

Boin et al. (2005, p. 116) differentiate between 'puzzling' and 'powering' in post-crisis learning. The former is about finding answers to such questions as what went wrong, why and what should be done to avoid the same incident in the future. The latter is then about the capacity to make substantial changes. In general, the post-crisis settlement provides an opportunity for new political initiatives and competence between different policy lines. A crisis may even open up an opportunity for drastic policy changes that have little to do with the actual crisis lessons. This is what is sometimes called management by crisis instead of crisis management. Drennan, McConnell and Stark describe this as "the political reality of accountability" (2015, pp. 192, 210), a process whereby decision-makers have to explain and defend, and often post-factum rationalise, their actions related to the crisis, and commit themselves to making changes when and where needed. This process will often include blame games, oftentimes encouraged or even organised by the media. Heroes, villains, escapologists or survivors will emerge from this after-game. Several issues, such as the nature and sector of the crisis, timing, political situation, interdependencies and symbolic value of the crisis are among those that affect this crisis aftermath.

The edited volume by Boin, McConnell and 't Hart (2008) on governing after a crisis is a thoroughly political rather than managerial account of crises, representing what was referred to above as the political science school of thought in post-crisis learning literature. It includes five case studies on blame games after a crisis, which are worth reviewing to draw a more detailed picture of this phenomenon in post-crisis management. Preston's (2008) account of President George W. Bush's crisis management of Hurricane Katrina in 2004 concludes that it was the President's own fault that he lost the blame game. This was mainly due to his leadership style, lack of engagement and the closed nature of his advisory system. Similarly, Olmeda (2008) argues that the Spanish government lost the 'meaning-making battle' to the opposition by initially providing the wrong interpretation of the terror attacks in Madrid in 2004. Consequently, they lost the subsequent elections. This, Olmeda (2008, p. 83) says, shows how "a crisis can create a window of opportunity for political entrepreneurs to advance a radically different frame and reap the political spoils of winning such a framing contest". Bytzek's (2008) narrative on Chancellor Gerhard Schröder's handling of the 2002 Elbe flood in Germany, on the other hand, demonstrates how the operational capacity, but principally the symbolic component, of crisis management contributed in helping Schröder's coalition government to retain its majority in the elections later the same year against previous odds.

An interesting comparative study is Brändström, Kuipers and Daléus's (2008) chapter, which deals with the patterns of blame management in Scandinavia, especially in Finland, Norway and Sweden, concerning the politics of the Boxing Day tsunami in East Asia in 2004. Their puzzle is why the same type of crisis, which resulted in the deaths of hundreds of citizens in these highly similar polities with their fairly similar administrations, nonetheless triggered markedly different political responses. In the authors' account, Finland and Norway managed rather well, whereas the Swedish post-crisis management was nothing short of a fiasco. This was due to many factors. The Finnish and Norwegian governments were quick to apologise for all the mistakes they had made in the turmoil surrounding the crisis. They had a better organised crisis management structure at home and even had a better crisis management capacity in place in the affected area. By contrast, the Swedish government continued to deny responsibility until the investigation commission finally stated that the prime minister had the overall responsibility for the shortcomings.

The last of the blame-game cases is Staelraeve and 't Hart's (2008) comparison between two Belgian cases, the so-called Dutroux case of two kidnapped and subsequently killed girls in 1996, and the contamination of poultry and eggs crisis of 1999. The authors refer to the aftermath of these two crises as 'exercises in political blame management'. While the first crisis was much more severe, the governing elite survived it much better than the second crisis. The authors end by emphasising the differences in such issues as timing, the breadth of the investigation scope, the investigation committee leadership and the media interest.

To sum up, the edited volume concludes that crisis management, following Beck's (1992) prophecy about a risk society, "has gone beyond the essentially low-level, technocratic sphere" that it perhaps was before (Boin, McConnell and 't Hart, 2008, p. 313).

References

Argyris, C. and Schön, D. (1978) *Organizational Learning: A Theory of Action Perspective*. Reading, MA: Addison-Wesley.

Beck, U. (1992) *Risk Society: Towards a New Modernity*. London: Sage Publications.

Bennett, D., Chiang, C.-F. and Malani, A. (2011) Learning During a Crisis: The SARS Epidemic in Taiwan. Working Paper 16955. National Bureau of Economic Research. Cambridge, MA. Available at: http://www.nber.org/papers/w16955

Birkland, T.A. (2006) *Lessons from Disaster: Policy Change after Catastrophic Events*. Washington, DC: Georgetown University Press.

Birkland, T.A. (2009) Disasters, Lessons Learned, and Fantasy Documents. *Journal of Contingencies and Crisis Management*, 17(3), pp. 146–156.

Boin, A. et al. (2005) *The Politics of Crisis Management: Public Leadership under Pressure*. Cambridge: Cambridge University Press.

Boin, A., McConnell, A. and 't Hart, P. (2008) *Governing After Crisis: The Politics of Investigation, Accountability and Learning*. Cambridge: Cambridge University Press.

Boin, A., 't Hart, P. and McConnell, A. (2009) Crisis Exploitation: Political and Policy Impacts of Framing Contests. *Journal of European Public Policy*, 16(1), pp. 81–106.

Borella, J. and Eriksson, K. (2013) Learning Effectiveness of Discussion-Based Crisis Management Exercises. *International Journal of Disaster Risk Reduction*, 5, pp. 28–37.

Borodzicz, E. and Van Haperen, K. (2002) Individual and Group Learning in Crisis Simulations. *Journal of Contingencies and Crisis Management*, 10(3), pp. 139–147.

Brändström, A., Kuipers, A. and Daléus, P. (2008) The Politics of Tsunami Responses: Comparing Patterns of Blame Management in Scandinavia. In Boin, A., Mcconnell, A. and 't Hart, P. (eds) *Governing After Crisis: The Politics of Investigation, Accountability and Learning*. Cambridge: Cambridge University Press, pp. 114–147.

Broekema, W. (2016) Crisis-Induced Learning and Issue Politicization in the EU: The Braer, Sea Empress, Erika, and Prestige Oil Spill Disasters. *Public Administration*, 94(2), pp. 381–398.

Bytzek, E. (2008) Flood Response and Political Survival: Gerhard Schröder and the 2002 Elbe Flood in Germany. In Boin, A., McConnell, A. and 't Hart, P. (2008) *Governing After Crisis: The Politics of Investigation, Accountability and Learning*. Cambridge: Cambridge University Press, pp. 85–113.

Carmeli, A. and Schaubroeck, J. (2008) Organisational Crisis-Preparedness: The Importance of Learning from Failures. *Long-Range Planning*, 41, pp. 177–196.

Carrel, L.F. (2000) Training Civil Servants for Crisis Management. *Journal of Contingencies and Crisis Management*, 8(4), pp. 192–196.

Cheng, S.S. and Seeger, M.W. (2012) Lessons Learned from Organizational Crisis: Business Ethics and Corporate Communication. *International Journal of Business and Management*, 7(12), pp. 74–86.

CSS. (2016) *Learning from Disaster Events and Exercises in Civil Protection Organizations*. Risk and Resilience Team, Center for Security Studies, Zurich: ETH Zurich.

Dawes, S.S., Cresswell, A.M. and Cahan, B.B. (2004) Learning from Crisis: Lessons in Human and Information Infrastructure from the World Trade Center Response. *Social Science Computer Review*, 22(1), pp. 52–66.

Deverell, E. (2009) Crises as Learning Triggers: Exploring a Conceptual Framework of Crisis-Induced Learning. *Journal of Contingencies and Crisis Management*, 17(3), pp. 179–188.

Deverell, E. (2015) Att lära avkriserfarenheter. In Deverell, E., Hansén, D. and Olsson, E.-K. (eds) *Perspektiv på krishantering*. Lund: Studentlitteratur, pp. 195–217.

Drennan, L., McConnell, A. and Stark, A. (2015) *Risk and Crisis Management in the Public Sector*. Second edn. New York: Routledge.

Drupsteen, L. and Guldenmund, F.W. (2014) What Is Learning? A Review of the Safety Literature to Define Learning from Incidents, Accidents and Disasters. *Journal of Contingencies and Crisis Management*, 22(2), pp. 81–96.

Elliott, D. (2009) The Failure of Organizational Learning from Crisis – A Matter of Life and Death? *Journal of Contingencies and Crisis Management*, 17(3), pp. 157–168.

Fabbri, A.G. and Chung, C.J. (2009) Training Decision-Makers in Hazard Spatial Prediction and Risk Assessment: Ideas, Tools, Strategies and Challenges. In Duncan, K. and Brebbia, C.A. (eds) *Disaster Management and Human Health Risk*. Boston: WIT Press, pp. 285–308.

Ferreira Pedroso, F., et al. (2015) *Input Paper. Prepared for the 2015 Global Assessment Report on Disaster Risk Reduction. Post-Disaster Challenges and Opportunities: Lessons from the 2011 Christchurch Earthquake and Great Eastern Japan Earthquake and Tsunami*. UNISDR and GAR.

Government of Indonesia and United Nations. (2005) *Post-Tsunami Lessons Learned and Best Practices Workshop Report and Working Groups Output*. Jakarta, Indonesia.

Government of Sri Lanka and United Nations. (2005) *National Post-Tsunami Lessons Learned and Best Practices Workshop*. Colombo, Sri Lanka.

Grosjean, P., Ricka, F. and Senik, C. (2013) Learning, Political Attitudes and Crises: Lessons from Transition Countries. *Journal of Comparative Economics*, 41, pp. 490–505.

Grunnan, T. and Maal, M. (2015) Lessons Learned and Best Practices from Crisis Management of Selected Natural Disasters: Elicit to Learn Crucial Post-Crisis Lessons. Norwegian Defence Research Establishment. FFI-rapport.

Hällgren, M. and Wilson, T.L. (2011) Opportunities for Learning from Crises in Projects. *International Journal of Managing Projects in Business*, 4(2), pp. 196–217.

Hansén, D. (2008) The 1975 Stockholm Embassy Seizure: Crisis and the Absence of Reform. In Boin, A., McConnell, A. and 't Hart, P. (2008) *Governing After Crisis: The Politics of Investigation, Accountability and Learning*. Cambridge: Cambridge University Press, pp. 183–207.

Hansén, D. (2009) Effects of Buzzwords on Experiential Learning: The Swedish Case of 'Shared Situation Awareness'. *Journal of Contingencies and Crisis Management*, 17(3), pp. 169–178.

Hansén, D. (2015) Kriser och policyförändring. In Deverell, E., Hansén, D. and Olsson, E.-K. (eds) *Perspektiv på krishantering*. Lund: Studentlitteratur, pp. 171–194.

Harvard Business School (2004) *Crisis Management: Master the Skills to Prevent Disasters*. Boston, MA: Harvard Business School Press.

Johansson, M. (2016) Data och lärande efter katastrofer. In Baez, U. and Becker, P. (eds) *Katastrofriskreducering. Perspektiv, praktik, potential*. Lund: Studentlitteratur.

Kamkhaji, J.C. and Radaelli, C.M. (2017) Crisis, Learning and Policy Change in the European Union. *Journal of European Public Policy*, 24(5), pp. 714–734.

Kayes, C. (2015) *Organizational Resilience: How Learning Sustains Organizations in Crisis, Disaster, and Breakdown*. Oxford: Oxford University Press.

Lagadec, P. (1997) Learning Processes for Crisis Management in Complex Organizations. *Journal of Contingencies and Crisis Management*, 5(1), pp. 24–31.

Levy, J.S. (1994) Learning and Foreign Policy: Sweeping a Conceptual Minefield. *International Organization*, 48(2), pp. 279–312.

McConnell, A. (2011) Success? Failure? Something In-Between? A Framework for Evaluating Crisis Management. *Policy and Society*, 30(2), pp. 63–76.

Milton, N. (2010) *The Lessons Learned Handbook: Practical Approaches to Learning from Experience*. Oxford: Chandos.

Moteff, J.D. (2012) *Critical Infrastructure Resilience: The Evolution of Policy and Programs and Issues for Congress*. August 23, Congressional Research Service 7–5700. Available at: http://www.fas.org/sgp/crs/homesec/R42683.pdf

Moynihan, D.P. (2008) Learning Under Uncertainty: Networks in Crisis Management. *Public Administration Review*, 68(2), pp. 350–365.

Moynihan, D.P. (2009) From Intercrisis to Intracrisis Learning. *Journal of Contingencies and Crisis Management*, 17(3), pp. 189–198.

Müller-Seitz, G. and Macpherson, A. (2014) Learning During Crisis as a 'War for Meaning': The Case of the German Escherichia Coli Outbreak in 2011. *Management Learning*, 45(5), pp. 593–608.

Olmeda, J.A. (2008) A Reversal of Fortune: Blame Games and Framing Contests After the 3/11 Terrorist Attacks in Madrid. In Boin, A., McConnell, A. and 't Hart, P. (eds) *Governing After Crisis: The Politics of Investigation, Accountability and Learning*. Cambridge: Cambridge University Press, pp. 62–84.

Preston, T. (2008) Weathering the Politics of Responsibility and Blame: The Bush Administration and Its Response to Hurricane Katrina. In Boin, A., McConnell, A. and 't Hart, P. (eds) *Governing After Crisis: The Politics of Investigation, Accountability and Learning*. Cambridge: Cambridge University Press, pp. 33–61.

Rerup, C. (2009) Attentional Triangulation: Learning from Unexpected Rare Crises. *Organization Science*, 20(5), pp. 876–893.

Roux-Dufort, C. and Metais, E. (1999) Building Core Competencies in Crisis Management Through Organizational Learning: The Case of the French Nuclear Power Producer. *Technological Forecasting and Social Change*, 1, pp. 13–127.

Schiffino, N. et al. (2017) Postcrisis Learning in Public Agencies: What Do We Learn from Both Actors and Institutions? *PolicyStudies*, 38(1), pp. 59–75.

Shaw, R. (2014) Post-Disaster Recovery: Issues and Challenges. In Shaw, R. (ed.) *Disaster Risk Reduction: Methods, Approaches and Practices*. Tokyo: Springer, pp. 1–13.

Sommer, M. (2015) Learning in Emergency Work. PhD thesis UiS no. 275, Stavanger: University of Stavanger.

Staelraeve, S. and 't Hart, P. (2008) Dutroux and Dioxin: Crisis Investigations, Elite Accountability and Institutional Reform in Belgium. In Boin, A., McConnell, A. and 't Hart, P. (eds) *Governing After Crisis: The Politics of Investigation, Accountability and Learning*. Cambridge: Cambridge University Press, pp. 148–179.

Stead, E. and Smallman, C. (1999) Understanding Business Failure: Learning and Un-learning Lessons from Industrial Crises. *Journal of Contingencies and Crisis Management*, 7(1), pp. 1–18.

Sterman, J.D. (2006) Learning from Evidence in a Complex World. *American Journal of Public Health*, 96(3), pp. 505–514.

Stern, E. (1997) Crisis and Learning: A Conceptual Balance Sheet. *Journal of Contingencies and Crisis Management*, 5(2), pp. 69–86.

Turner, B.A. (1976) The Organizational and Interorganizational Development of Disasters. *Administrative Science Quarterly*, 21(3), pp. 378–397.

UK Cabinet Office. (2013) *Emergency Response and Recovery. Non-statutory Guidance Accompanying the Civil Contingencies Act 2004*. Civil Contingencies Secretariat. Available at: https://www.gov.uk/government/uploads/system/uploads/attachment_data/file/253488/Emergency_Response_and_Recovery_5th_edition_October_2013.pdf

UNISDR. (2009) UNISDR Terminology on Disaster Risk Reduction. United Nations International Strategy for Disaster Reduction (UNISDR), Geneva, Switzerland, May 2009. Available at: http://www.unisdr.org/we/inform/terminology

Veil, S.R. (2011) Mindful Learning in Crisis Management. *International Journal of Business Communication*, 48(2), pp. 116–147.

Westrum, R. (1996) Human Factors Experts Beginning to Focus on Organizational Factors in Safety. *ICAO Journal*, 51(8), pp. 6–8, 26–27.

Westrum, R. (2004) A Typology of Organisational Cultures. *Quality and Safety in Health Care*, 13(Suppl 2), pp. 22–27.

8 Conclusions

This book is drawing to a close. It has discussed the crisis management cycle in some detail, focusing on one phase in each chapter. By definition, this cycle is a continuous one, and in practice the phases are also overlapping, not strictly sequential crisis management activities. The aim of this book has been to use the crisis management cycle in order to draw a holistic and generic picture of the field of crisis management, on the one hand, and to identify the competing, overlapping or co-existing discourses, debates and issues within this field of study, on the other. A tabular summary of the results of this exercise will be useful at this juncture (see Table 8.1 at end of chapter).

8.1 Summary of the main observations

When it comes to **Risk assessment**, explained in Chapter 2, the current dominant discourse is that of the International Organization for Standardization (ISO) 31000 standard. This is an approach to identifying risks, analysing their likelihood and impacts, and selecting the risks to be treated. The standard is mainly applied in organisations as a basic set of adaptable guidelines, but recently one can find its application at the societal level when assessing societal risks, especially in the field of safety and security. There is no explicitly articulated theory behind the standard, but there are standardised definitions and terminology, and a process-based typological structure that is helpful when planning the practical work. The methodological approaches grouped under this structure are rather detailed, including a range of more or less familiar qualitative, semi-quantitative and quantitative techniques. When it comes to safety engineering in particular, the quantitative techniques benefit from rather sophisticated mathematical models.

If ISO 31000 is the gold standard when it comes to risk management, the concept of risk governance has emerged to oppose the management approach (even before ISO 31000 was adopted), arguing that it is too technological, and assuming that risks can be objectively identified, analysed and evaluated. Risks in this discourse are understood to be more like social constructions than any objective reality. Hence, this discourse is more sensitive to risk perception. Its unit of analysis is society at large, or even the global society, both in terms of horizontal and vertical coordination, as well as the negotiation processes between different stakeholder groups. The discourse clearly arose from the more generic political science multi-level governance discourse, and uses qualitative social science and policy analysis methods to analyse the governance process. More recently, one can identify some attempts to combine the risk management and risk governance discourses, which would, however, presuppose the selection of a common unit and level of analysis.

Chapter 3 on **Prevention** started with a discussion on safety culture as a precondition for prevention and mitigation. This highly normative discourse clearly focuses on organisations, which can be differentiated in relation to the variable that determines how safety and security issues, or risks, are embedded in the organisational culture and practices. While no well-developed theory seems to exist even here, a general understanding has been reached on the basic elements of safety as a subculture within the organisational culture. This duly provides a point of comparison or benchmark when conducting empirical research. Conducting research on safety culture usually focuses on actual organisations, and the methods include surveys, interviews, statistics and observations, among others.

If safety culture discourse is about the foundations of prevention, other discourse deals more concretely with prevention and mitigation strategies. This discourse is best framed by relying once again on the ISO 31000 standard. In accordance with this standard, risk treatment follows risk assessment. This in turn is divided into several options or strategies, such as avoiding the risk, changing its likelihood or consequences or sharing the risk, to name just a few. These are designed to be applied individually, but most often in the appropriate combinations. Again, no clearly articulated theory can be found despite well-standardised definitions and terminology, as well as a typological structure for considering the risk-treatment options. As essential financial and other costs are involved in selecting the risk-treatment strategies, this is usually no mean feat, often calling for sophisticated qualitative, semi-quantitative and quantitative evaluations and calculations.

In Chapter 4 on **Preparedness**, one cannot identify clear discourses but rather some central themes that form their own sub-sections in the literature. The first concentrates on planning and organising preparedness. Its normative advice is to make preparedness planning part of the strategy in any organisation. There is no theory underpinning this advice, but there are some normative typologies, guidelines and best practices that can be used as a basis for developing preparedness planning in practice. In more academic but often descriptive studies, the methodology is often a post-crisis qualitative case study pointing out the failures of preparedness planning against the generally accepted best practices.

A perhaps more concrete and research-oriented part of the literature discusses the individual elements of preparedness, usually focusing on only one at a time, such as capacity- or capability-building. The unit of analysis is mostly an organisation or a professional branch (such as rescue personnel or political decision-makers). While there does not appear to be any clearly articulated theoretical background to this literature, relying as it does on certain normative guidelines and best practices, in its more sophisticated versions one can find testing procedures related to the functionality of different training methods, for instance, vis-à-vis the target groups and their professional needs.

Another and somewhat separate theme within preparedness is early warning. The main issue here is how to organise it in the best possible way. The unit of analysis is usually an organisation which, in 'high politics' crises, is the government. The theoretical, conceptual and methodological approaches range from typical social science case studies to highly technological approaches, such as complex simulations.

Response is discussed in Chapter 5 in terms of four discourses or debates. The first of these concerns framing a crisis. While some situations are crises by definition, a crisis can sometimes be 'constructed'. Even in a clear crisis situation, such as a natural disaster, the crisis can be framed differently. This discourse focuses mostly on political decision-makers, and sometimes also on organisations. A certain amount of literature is dedicated to crisis framing, most notably under the related concept of sensemaking. The methodology is normally rather descriptive policy analysis based on case studies.

The second discourse is rather practical-normative and discusses the issue of how to organise response. It builds on rivalry between such approaches as centralisation versus decentralisation, or command and control versus horizontal coordination. The current trend is generally to defend the latter against the former. This literature touches on political leadership and organisations in general, but most notably concrete crisis management in organisations. Typically, if there is any theoretical background, the literature is rooted in organisational studies or administrative and political science. The methodologies used are qualitative in nature, often using single or small-n case studies.

The third discourse, or rather debate between rival schools of thought, is perhaps the most theoretical in the crisis management literature. It endeavours to explain crisis decision-making, normally only post factum and mostly focusing on political leadership, but also applicable at the organisation level. The disciplinary roots of this debate are most notably in international relations studies, which in turn draw on economics, organisation studies and psychology. The theoretical background is strong, including such established schools as rational choice, bounded rationality, bureaucratic policy, organisational and institutional theories, and a variety of psychological theories. These types of studies mostly use qualitative methodologies applied to case studies, with the exception of rational choice theory, which usually uses game theory as the main methodological tool.

The last issue discussed under the response phase was crisis communication. This does not form a coherent school, discourse or theme as such. Crisis communication is defined in different ways, and it is traditional to focus on just one of its dimensions. Chapter 5 emphasises both external and internal communication, as well as so-called network communication in complex crises. The normative issue is how this communication should be organised. The literature focuses on the crisis decision-makers, as well as the internal and external stakeholders involved. Organisation studies and media studies provide some conceptual background for this literature. In research-oriented applications, qualitative case studies using interviews as a source of data collection are common.

Chapter 6 on ***Recovery*** is the shortest in the current book, an outcome that is perhaps reflective of the fact that much less has been written on this theme compared to the other crisis management phases. Most of the recovery literature centres on what this book refers to as recovery planning. Hence, the issue is not about the actual recovery phase, but rather about preparedness, identifying the elements that have to be in place in order to facilitate post-crisis recovery. There is no real theory behind this discourse, but it is explicitly normative and tends to rely on, or develop, best practices and lessons learned. The methodological solutions are most often qualitative, with some statistics added, and case studies are used as the source of evidence.

The second and more recent discourse one can identify concerns resilience. While resilience is a wider concept than recovery, its central element is the issue of how systems can quickly recover after they have failed. In this book, the societal, organisational, technological, economic and psychological resilience domains are briefly discussed in terms of recovery. The unit or level of analysis varies depending on the domain that is under discussion. In this sense, the analysis can focus, for instance, on the individual, facility, organisation, market or supply chain, or society at large. Theoretically, too, the conceptual framework depends on the domain in focus, from social science approaches to engineering. One can find examples of the use of qualitative, semi-quantitative and quantitative methodologies in this discourse.

Finally, Chapter 7 on post-crisis ***Learning*** identified two discourses or debates stemming from different disciplinary origins: organisation studies and political science. They nevertheless share the same general research questions, namely, identifying the mechanisms, conditions and reasons that are entailed when society or organisations learn, or fail to learn, from a crisis experience. A prerequisite for both discourses is an awareness that post-crisis learning is essentially different from 'normal' learning. The organisational approach naturally focuses on organisational learning, but because individual learning can be said to precede organisational learning, this approach also looks at the cognitive aspects involved. The type of learning in question is also discussed here, differentiating between mere feedback and a deeper level of learning. Different types of qualitative analysis, often focusing on case studies, form the main methodological approach.

Political science studies, for their part, usually focus on political leadership. The germ of a theory can be found in the background to this literature, as a typological framework has emerged for understanding the process of learning, including such elements as using a crisis as an opportunity to open policy windows, and the blame games that ensue after a crisis. Some testable hypotheses have been developed within this discourse. Conceptually, this literature is rather typical of political science in that it uses qualitative approaches and focuses on case studies.

8.2 From practice to theory?

There is an anecdote about the General Secretary of the Communist Party, Nikita Khrushchev, which claims that during his first visit to the United States, his host, President John F. Kennedy, was proudly showing his Soviet guest all the miracles of the capitalist economic system when Khrushchev merely quipped: "Yes, it looks to work in practice, but I doubt whether it would work in theory."

By the same token, the field of crisis management seems to be a field that is rather developed and an everyday phenomenon in its different phases in practice, but not in theory. Crisis management as a field is, by definition, characterised by a close relationship between the political and practical agenda. This is also reflected in the literature, a substantial or perhaps even major part of it comprising practical-normative reports, descriptive guidelines and best practices.

While there is no holistic theory about crisis management, there is, instead, a collection of mid-level more or less theoretical treatments of its different elements, originating from a range of disciplines. With just a few exceptions, these discourses rarely communicate with each other, but rather develop in isolation. The result is a field that is highly fragmented, with each new middle-range theoretical avenue just adding to this fragmentation. This kind of field has no clear identity and boundaries, which can be beneficial in terms of innovation, but is less so when it comes to creating a discursive community. Indeed, there is conceptual and discursive chaos in this respect, without any theoretical or methodological cohesion. This can also be a good thing, but only on the condition that some kind of core exists that would allow fruitful debate and communication.

Many of the challenges stem from the fact that crisis management is not 'owned' by any discipline in particular. There are basically three possibilities for the way in which a field that is of interest to many academic disciplines can organise this diversity of theoretical and methodological commitments. First, this can be handled in terms of multidisciplinarity. This means that several disciplines contribute to one field or one set of problems from their own specific starting points, but do not genuinely communicate with each other. One might conclude from the current book's review that this is the state of the art in crisis management studies today. Second, it is possible to create an interdisciplinary field, which would imply that the field would cross disciplinary borders. In practice, this would mean that an engineer and a social scientist, for instance, would co-author a study, combining their knowledge and disciplinary backgrounds, thereby creating a new way to approach a problem. This type of interdisciplinarity can be found in the crisis management literature, but only rarely. Third, this development may lead to transdisciplinarity, that is, developing a new holistic discipline that would demand skills from many disciplines from its scholars. It seems that this sort of development is a long way off in the crisis management field, if indeed it ever materialises at all.

This leaves us with a puzzle about the kind of theoretical level that is actually possible in the field of crisis management, and which would go beyond mid-level theorising on its individual elements. It is quite clear that it is too ambitious to aim for a completely explanatory theory of crisis management in the positivist theory-of-science sense, that is, not only explaining but also predicting phenomena. However, the crisis management cycle provides, as does any descriptive theory, concepts and taxonomies that reveal something about the characteristics of the phenomenon and its elements. Moreover, it also tells us something about the way in which these different elements are related to each other. This is the level of theorising that is currently achievable in crisis management studies. This book has striven to make inroads in that direction, and has hopefully managed to put the multiple pieces in the crisis management puzzle into some sort of logical order that will point the way to a more consolidated field.

Table 8.1 Crisis management discourses and debates

REPRESENTATIVE EXAMPLES/CRISIS MANAGEMENT CYCLE PHASES	Main discourses, approaches and debates	Main issues and research problems	Unit/level of analysis	Conceptual framework	Methodological approaches
Risk assessment	Risk assessment (ISO 31000)	Identifying risks, analysing their likelihood and impacts, selecting the risks to be treated	Organisation, society	Standardised vocabulary, process structure	Qualitative, semi-quantitative and quantitative techniques
	Risk governance	Hierarchical and horizontal coordination of risk perceptions	Society, global society	Multilevel governance discourse	Qualitative social science and policy analysis
Prevention	Safety/organisational culture	Building an organisation where safety and security issues, or risks, are embedded in the organisational practices	Organisation	Generally agreed upon understanding of basic elements of safety / organisational culture	Surveys, interviews, observation
	Risk treatment (ISO 31000)	Systematically considering risk treatment options for the identified risks and using them in appropriate combinations	Organisation, society	Standardised vocabulary, standardised typology of the risk-treatment options	Qualitative, semi-quantitative and quantitative evaluations, and calculation on the pros and cons of the treatment options

REPRESENTATIVE EXAMPLES/CRISIS MANAGEMENT CYCLE PHASES	Main discourses, approaches and debates	Main issues and research problems	Unit/level of analysis	Conceptual framework	Methodological approaches
Preparedness	Planning and organisation	Making crisis preparedness a normal task of any organisation	Organisation	Normative standard and best practices	Often post-crisis qualitative case studies pointing out the failures in preparedness
	Elements of preparedness	Capacity- and capability-building	Organisation, professional branch	Normative guidelines and best practices	Typological qualitative analysis, but also some testing of approaches in terms of their functionality
	Early warning	How to organise crisis early warning	Organisation, government	Social science and technological approaches	From qualitative to highly sophisticated technological methodologies, case studies

REPRESENTATIVE EXAMPLES/CRISIS MANAGEMENT CYCLE PHASES	Main discourses, approaches and debates	Main issues and research problems	Unit/level of analysis	Conceptual framework	Methodological approaches
Response	Framing of a crisis	How a crisis is identified and framed; could it have been framed differently?	Political leadership, organisation	Debates about sensemaking	Qualitative analysis
	Organisation of crisis decision-making	Centralisation vs. decentralisation; command and control vs. horizontal coordination	Political leadership, organisation, concrete crisis management in organisations	Organisational studies, political science	Qualitative analysis, often through case studies
	Explaining crisis decision-making	Which theory explains crisis decision-making?	Political leadership, organisation	Rational choice, bounded rationality, bureaucratic policy, organisational and institutional theories, psychological theories	Mostly qualitative methodologies applied to case studies; rational choice game theory
	Crisis communication	How should crisis communication be organised?	Crisis decision-makers internal and external stakeholders	Organisational studies, media studies	Qualitative analysis, case studies, interviews

REPRESENTATIVE EXAMPLES/CRISIS MANAGEMENT CYCLE PHASES	Main discourses, approaches and debates	Main issues and research problems	Unit/level of analysis	Conceptual framework	Methodological approaches
Recovery	Recovery as planning	Identifying the elements that have to be in place in order to facilitate post-crisis recovery	Organisation, community	Descriptive and normative best practices	Mainly qualitative analysis, some statistics, case studies
	Resilience discource (societal/community, organisational, technological, economic, psychological)	Analysing and enhancing the ability of a system, community or society exposed to hazards to (resist, absorb, accommodate to and) recover from the effects of a hazard in a timely and efficient manner	Individual facility, organisation, market/supply chain, society, regulators	A variety of approaches, depending on the application area, from social sciences to engineering	Qualitative, semi-quantitative and quantitative methodologies
Learning	Organisation studies	Identifying the conditions under which society or organisations learn or fail to learn from a crisis experience	Organisation, individual	Cognitive and institutional analysis	Qualitative methodologies, case studies
	Political studies		Political leadership	Policy and conceptual analysis	

Index

absorption 132–3
Acar, C. 120
acceptance of risk 23–4
accident, normal 18, 43
active redundancy 83–4
ad hoc groups 77
adaptive learning 150–1
Agreement on Cooperation on Marine Oil Pollution 84
agreements 84–5, 101
Alexander, D. 73, 129
Allison, G.T. 78–9, 103, 107–9
amplification, social 26
An, C. 35–6
AND gate 27
Andersen, T.J. 62
Anderson, S. 25
Andersson, J. 58
ANSI/ASIS 136
Appropriate Level of Protection (ALOP) 59
Arctic Council 84
Argyris, C. 152
As Low As Reasonably Practicable (ALARP) 23, 59–61
attentional tunnelling 52
Aurora Borealis 30
Austina, L.L. 128
Australia 159–60
aviation 42–3, 52
avoidance of risk 44–9

backup 82–3
Baltic Sea 49
barriers 50–1, 53–5, 147
Beck, U. 18–19, 165
Belgium 165
Bellavita, C. 18
Bennett, D. 161
Benoit, W.I. 117

Berlin, J.M. 87
Bernstein, P. 10
Best Available Technology (BAT) 59
Bharosa, N. 116
bias 106–7, 109–11
Birkland, T.A. 149, 152, 160, 163
blame games 164–5
Boin, A. 3–4; learning 151, 156–7, 163–4; preparedness 69, 71; response 98–9, 115, 117
Boiral, O. 19
Boland, P.J. 84
Borell, J. 87
bounded rationality 78, 105–7
Bow-tie Analysis 27, 49–50
Boxing Day tsunami 163–4
brainstorming 30–1
Brändström, A. 164
Brecher, M. 103
Broekma, W. 162–3
Bruneau, M. 138
Bullock, J.A. 50, 70, 128, 134
burden of proof 46–7
bureaucratic organisation 42
bureaucratic politics 105, 107–8
Burton, C.G. 135
Bush, G.W. 164
business 2; assessment 10–11, 15, 22–3; learning 158–60; prevention 59, 62; recovery 128, 130, 136; response 102, 104, 117–18, 121
business continuity 130, 136–7
business impact analysis (BIA) 130–1
buzzwords 160
Bynander, F. 115
Bytzek, E. 164

capability-building 85–8
capacity-building 79–81

Index

Carlström, E.D. 87
Carmeli, A. 71, 148, 151, 153
Carrington flare 30
cascading effects 9, 15, 20–1, 30, 50
Cauberghe, V. 117
causation 30, 82–3, 113
centralisation 76–8, 100, 102, 129, 157
change 88, 150–8, 160
chaos theory 18, 139
Cheng, S.S. 159
Chernobyl 40–1, 51, 153
Chiang, C.-F. 161
child safety 48
China 76, 129
civil emergencies 1–2, 33, 106, 116, 119
civil protection 11, 15, 22, 33–4, 77, 85, 149
Civil Protection Mechanism 89
Claeys, A-S. 117
climate change 112
Cockpit Resource Management (CRM) 43
cognition 97–8, 105, 109–11
cognitive tunnelling 52
collaboration 84–5, 87, 115, *see also* coordination
collegiate leadership 102–3
command and control 76–8, 100, 102, 115
communication 26, 111–21, 161, 171–2; early warning as 112–15; external 116–20, 159; internal 120–1; network/coordination 115–16
communicative planning 72–3
community participation 129–30
community resilience 134–6
companies *see* business
competitive leadership 102
complexity 18, 30, 43; complex disasters 112; complex learning 152, 158
compulsive personality 103
conceptual knowledge 149–50
Condition Based Maintenance (CBM) 54–5
consequences 12, 20–3, 29, 33–5, 80–1; reducing 49–50, 53
context 15–16
contingency planning 72–6, 84, 86, 101
contingent learning 150
control and command 76–8, 100, 102, 115
Coombs, W.T. 117
coordination 115–16, 130, *see also* collaboration

Copenhagen School 98
Coppola, D.P. 50, 70, 128, 134
corporations *see* business
corrective maintenance 54–5
cost–benefit analysis 24, 61, 111
counterfactuals 82
Cox, A.L. 35
creeping crises 97, 113
Crew Resource Management (CRM) 43
Crimea 111
crisis 2–3; decision-making theories 103–11; identifying/framing 97–100; life cycle 4–5; managing by 102, 163–5, *see also* communication
Crisis Communication Team 118
crisis management 1–4, 165; definitions 3–4; from practice to theory? 173–4
crisis management cycle 4–6, 146, 169–73
criteria 15–16, 33
critical infrastructure 21, 55–8, 74, 80–1, 131, 136–7
Critical Infrastructure Task Force 160
critique 43
crowdsourcing 120
CSS 158, 161
Cuban missile crisis 103, 107
culture, organisational 40–3, 88, 158, 170
'currently tolerated' approach 24
Cutter, S.L. 135
cyber security 21, 132

Daléus, P. 164
damage assessment 134
Darwinism 139
data 16–17, 20, 28–30, 112
data overload 52
Davis, I. 129
De Bruijne, M. 56–7, 131
decentralisation 77–8, 100, 102, 129
decision-making 4, 48, 100–2, 115, 130, 157; preparedness 72, 76–9; theories of 103–11
defining a situation as a crisis 97–100
definitions: crisis management 2–4; critical infrastructure 81; early warning 89; maintenance 54; preparedness 69–70; prevention 39; rationality 104; recovery 127–8; redundancy 82; resilience 132, 137–9; response 96; risk 11–15
denial 117–18, 159–60
DeRouen Jr., K. 110
descriptive theory 174
Deverell, E. 148, 150, 154, 156

Index

disasters 80–1, 160; complex 112; recovery 127–30, 134
disaster losses 19
disaster relief 96, *see also* response
disaster risk management 89–90
disaster risk reduction (DRR) 39, 114
discourses 9, 14, 18, 43, 62, 169–72
discussion-based exercises 86–7
'disease burden' approach 24
dispositions 149–50
DNV 22
domino causality 30
double-loop learning 152–3, 157–8, 160
Drennan, L. 50; learning 152–3, 162, 164; preparedness 70–2, 74–5; response 101–2, 115
drills 86, 162
Drupsteen, L. 158–9
Dugdale, J. 120
Dulek, R.E. 111
Dutroux case 165
Dynes, R. 73

early warning 51–4, 99, 110–11, 148, 171; as communication 112–16; as preparedness 88–90
earthquakes 31–2, 120, 138
ecology 134–5
economy 21–2, 24, 115, 138–9
education 86
efficiency 78
Egan, M. 56
El-Neweihi, E. 84
Elbe flood 164
electricity 22–3, 30, 70
elite escape 156–8
Elliott, D. 156, 162–3
emergencies 1–2, 53; assessment 18, 33; learning 152, 158; preparedness 73–4, 76, 80, 86–7, 90; recovery 128, 134; response 104, 106, 112–16, 119
emergency management cycle 4–5
emergent risks 29–30
employees 41–2, 120–1
Emrich, C.T. 135
Enander, A. 71
engineering 9, 81–4, 137
environment 21–2, 46
equality, principle of 76
equipment 79
ergonomics 52, 114, 153
Erikson, K. 87
Europe 55, 57, 77, 103

European Commission 11–12, 21–2, 29, 46–7, 85
European Standard EN13306:2001 54
European Union (EU) 41, 47, 58, 85, 89, 154
evaluation, post-crisis 161–3
event scenarios 34
Event Tree Analysis (ETA) 27–30
exercises 86–7, 148
experience-based learning 151
experts 30–1
explanation-based learning 151
exposure (E) 12

fast learning 152
Fault Tree Analysis (FTA) 27–9
Felici, M. 25
Fewtrell, L. 24
Fiksel, J. 82
financial crises 79, 140
financial risks 22, 45, 62
fine-tuning 152
Fink, S. 71, 79, 101
Finland 164–5
Fioritto, A. 58
Fisher Liub, B. 128
Ford, R. 134
foreign policy crises 110–11
Forsberg, T. 99, 111
framing a crisis 97–100
France 153
Frandsen, F. 121
Fukushima Daiichi 32, 48
full-scale exercises 86
functional relationships 12
fuzzy risk matrix 36

Garvey, M. 62
gender 25
generative organisation 42
Germany 48, 164
globalisation 3, 56
God(s) 10
Goldenmund, F.W. 158–9
Golnaraghi, M. 89–90
Google 12
governance approaches 9, 14, 18, 25–6, 62, 130, 169–70
government 2; assessment 11, 15; learning 154, 160–1; preparedness 71–2; prevention 55–8; recovery 129–30; response 114–15, 117–18, 121
Great Tohoku earthquake 120
groupthink 109–10, 158

182 Index

Grunnan, T. 162
Guha-Sapir, D. 140
guidewords 31
Guldenmund, F. 41–2

Haddow, G.D. 50, 70, 128, 134
Haiti 120
Hale, D.P. 111
Hale, J.E. 111
Hansén, D. 156–8, 160
Harvard Business School 155
Healthcare Failure Modes and Effects Analysis (HFMEA) 32
Hede, S. 71
Helsloot, I. 100
Hémond, Y. 69
Hermann, C.F. 2, 103
Hertz, S. 74
heuristic mode 25
hierarchy 100, 102, 115
high reliability organisation (HRO) 42–3, 88
Hirshbein, R. 97
historical analogies 97–8, 100
historical data 20, 29
Holladay, S.J. 117
Holmström, P. 131–2
Holsti, O. 103
Homeland Security Advisory Council 160
hospitals 80
HSSAI 132
Huihui, N. 35–6
human error 51–3, 153
human factors 51–3, 97–8, 113–14
human impacts 21–2
human resources 74
Hunter, P.R. 24
Hurley, D. 57–8
Hurricane Katrina 160, 164

identifying a crisis 97–100
imagination 16
impact analysis 12, 20–3, 33–5
inclusion 26
inertia, organisational 156–7
Information and Communication Technology (ICT) 45, 56, 58
information processing 25, 52, 110, 113, 116
information sharing 57–8, 116
infrastructure 21, 55–8, 74, 80–1, 131, 136–7
INSAG 40–1

inspection 55, 114
institutions 107–9, 114, 163, *see also* organisation
instrumental planning 72–3
instrumental rationality 104
insurance 10–11, 59, 159
integration 26
interdisciplinarity 173
International Atomic Energy Agency (IAEA) 40
International Federation of Red Cross 75
International Nuclear Safety Group 40–1
International Organization for Standardization *see* ISO
International Statistical Classification of Diseases and Related Health Problems 139
internet 119–20
intra-crisis learning 148–9, 153
ISO (International Organization for Standardization) 11–12, 39, 49
ISO-2800 74
ISO-9000 3–4
ISO-22301 74
ISO-28002 136
ISO-31000 147, 169–70; assessment 13–16, 23–5; prevention 39–40, 44, 48, 61–2
ISO-41000 74

Janis, I.L. 103
Janssen, M. 116
Japan 120
Jarman, A.M.G. 102
Jensen, L.-M. 74
Jin, Y. 128
Johansen, W. 121
Johansson, M. 153–4, 163
Jorritsma, J. 100

Kamkhaji, J.C. 148–50
Kaneberg, E. 74
Kanellopoulos, A. 140
Kapucu, N. 115
Kennedy, J.F. 173
Khrushchev, N. 173
Kim, S. 25
Klinke, A. 47
the 'knowable' 18
knowledge 149–50; management 74
Koeppinghoff, C. 120
Kouzmin, A. 102
Krausman, E. 139
Kreps, G. 73

Kriebel, D. 48
Kuipers, A. 164

Lagadec, P. 43, 69, 113
Lajksjö, Ö. 71
Lalonde, C. 19
Landstedt, J. 131–2
leadership 75–8, 100–3, 115, 137
learning 4, 9, 146–65, 172–3, 178; blame games 163–5; failure to learn 155–61; fast/slow 152–3; as horizontal challenge 147–9; is post-crisis learning different? 149–52; post-crisis evaluation 161–3; preparedness and 87–8; social 26, 161; who learns? 153–5
LEDDRA Project 135
Lee, J. 116
legislation 23, 45–8, 56, 90, 114
Leiden-Uppsala school 103
Levy, J.S. 147
liability 45, 56, 79
likelihood 12, 20–1, 34; reducing 49–50, 53
local level 129–30, *see also* decentralisation
logistics 85
London 10
Lunde, I.K. 70
Lussand, K. 151

Maal, M. 162
McConnell, A. 3; learning 152–3, 162, 164; preparedness 70–2, 74–5; response 98, 101–2, 115, 117
McDougall, A. 70
McEntire, D.A. 134
McManus, S. 136
Macpherson, A. 148
macro-drivers 34
Madrid 164
maintainability 138
maintenance 54–5; unplanned 138
Malani, A. 161
Malm, A. 58
management 3, 41–2, 79, 121; by crisis 102, 163–5
mathematics 10, 12, 20, 28
matrix 34–6
meaning-making 4, 100, 164, *see also* sensemaking
measurement 42
media 97, 116–21, 159, 164
megatrends 34
memory 97–8, 100, 158–9; requisite 52

Merriam-Webster 127
Mesopotamia 10
Metais, E. 153
methodologies 16, 20, 22, 26–32, 169–72
Mintz, A. 110
mitigation 39–40, 49–50, 170
Mitroff, C.M. 73
modularity 138
money 79, *see also* financial risks
monitoring 88
Moteff, J.D. 133, 160
Moynihan, D.P. 148, 153–4
Müller-Seitz, G. 148
multi-hazards 15–16
multi-risks 30
multidisciplinarity 14, 173
Muraki, Y. 120
Mussington, D. 57

narcissistic personality 103
NASA 82
NATECH 31
National Incident Management System 77
natural disasters *see* disasters
Netherlands 76
network communication 115–16
network learning 154–5
network management model 77, 115
Ning, C. 35–6
nonlinear dynamics 18, 30, 113
normal accident theory 18, 43
normative criteria 100
normative rules 104
Norway 22, 33, 75–6, 151, 164–5
nuclear power 153; assessment 24, 31–2; preparedness 76, 80; prevention 40–2, 48–9, 51

objectivity 25
Olmeda, J.A. 164
Operator Security Plan 58
OR gate 27
organisation 2, 54–5, 76–9, 81, 107, 114–16; internal communications 120–1; model 78; organisational culture 40–3, 88, 158, 170; organisational inertia 156–7; organisational learning 153–4; organisational recovery 136–7; organisational routines 105, 108–9; organisational trust 159
organisation studies 147
out-of-the-loop syndrome 52

Palme, O. 158
paradigm shift 152–3
parametric choice situation 106
paranoid personality 103
Parker, C.F. 115
Parnell, J.A. 72
participation, community 129–30
passive redundancy 83–4
path dependence 151
pathological organisation 42
Pearson, C.M. 73
perceptions: identifying/framing a crisis 97–9; of risk 24–6, 42, 45, 59, 61
performance, as resilience metric 133–4
performance loss triangle 132–3
Perrow, C. 18
personal resilience 139
personality types 102–3
Pinkowski, J. 72
Piotrowski, C. 72
planning 71–6, 170; prevention and 43–4; recovery as 128–31; response 101, 120–1, *see also* preparedness
police 151–2, 158
policy reform 152
poliheuristic decision-making 107
political science 147, 172–3
politics 21–2, 24, 102–3, 157; blame games 164–5; bureaucratic 105, 107–8; politicisation of crisis 162–3; use of history 97–8
pollution 84–5
polythink 110
positioning 118
Post, J.M. 103
post-traumatic stress disorder (PTSD) 139–40
Potts, H.W.W. 32
powering 163–4
practical-normative literature 100–3, 171
practice and theory 173
pre-arrangements 84–5, 101
precautionary principle 23, 45–8
'predefined probability' approach 24
preparedness 69–90, 170–1, 176; agreements/pre-arrangements 84–5; capability-building 85–8; capacity-building 79–81; early warning systems 88–90; learning and 148; organisation/procedures 76–9; planning 71–6; post-crisis evaluation 161; redundancy as 81–4; response and 100–1, 112, 120–1
Presidential Decision Directive 57
press conferences 119

Preston, T. 164
prevention 39–40, 170, 175; learning and 147–8; risk-taking as? 61–2
preventive maintenance 54–5
privatisation 55–8, *see also* business
probability 10–12, 19–20, 28–9, 34
procedural knowledge 149–50
procedures 78–9, 101, 108–9
proof, burden of 46–7
prospect theory 26, 109
protection *v.* resilience 131–2
psychology 106, 109–11, 139–40
'public acceptance' approach 24
public sector *see* government
public-private partnership (PPP) 55–8, 115
Purdy, G. 11
Pursiainen, C. 99, 111
puzzling 163–4

qualitative techniques 16, 20, 26–7, 29–32, 169–72
quantitative techniques 16, 20, 22, 26–9, 169–70
Quarantelli, E.L. 73

Radaelli, C.M. 148–50
Radvanovsky, R. 70
randomness 113
rationality 72–3, 78, 104–11, 171
readiness 69
reality 25
Reasonable Relationship 59
rebuild strategy 118
recoverability 134
recovery 9, 127, 172, 178; economic 138–9; learning and 149; organisational 136–7; as planning 128–31; psychological 139–40; resilience as 131–4; societal 134–6; technological 137–8
redundancy, as preparedness 81–4
reflection 26
refugee crisis 140
regulation 45–8, 55–8, 137
reinsurance 59
Reliability Centred Maintenance (RCM) 54
Renn, O. 19, 26, 47
reputation 117–20, 137, 160
requisite memory trap 52
Rerup, C. 148
resilience 43, 69–70, 80–2, 172; as recovery 127, 131–4

response 9, 96, 111, 171–2, 177; learning and 148–9; organising 100–3; post-crisis evaluation 161; recovery and 128, 134
responsibility, principle of 76
restoration 132–3
retaining risk 59
risk 11–12, 101; social amplification of 26
risk acceptance 23–4
risk analysis 14, 17, 19–23
risk assessment 9–36; conclusions 169–70, 175; context of 15–16; defining 12–15; learning and 147; prevention and 50–1, 54, 57; qualitative 29–32; recovery and 130–1; varying perceptions 24–6
risk aversion 25–6, 109
risk avoidance 44–9
Risk Based Inspection (RBI) 55
risk criteria 15–16, 33
risk evaluation 14, 16, 23–6
risk governance discourse 9, 14, 18, 25–6, 62, 169–70
risk identification 14, 16–20
risk management 13–14, 18, 39–40, 169–70
risk matrix 34–6
risk perception 24–6, 42, 45, 59, 61
risk reduction 39, 55–8
risk removal 48–9
risk retaining 59
risk scenarios 32–4
risk sharing 58–9
risk society discourse 9, 18, 43, 62, 165
risk taking 61–2, 109, 111
risk tolerance 23–4, 49, 59–60
risk treatment 39–40, 43–4, 49–50, 59–61
Robert, B. 69
Roberts, J. 103
Robinson, P. 57
Rodriguez-Llanes, J.M. 140
Roggi, O. 62
roles 107–8
Rose, A. 139
Rosenthal, U. 102
routines, organisational 105, 108–9
Roux-Dufort, C. 153
Runyan, R.C. 130
Russia 111
Ryu, Y. 25

safety 22, 61–2, 71, 129
safety culture 40–3, 114, 148, 153, 170

SARS 161
satisficing 106
Scandinavia 76, 78, 164–5
scenarios 32–4
Schauerbroeck, J. 71, 148, 151, 153
Schiffino, N. 147, 150
Scholtens, A. 100
Schön, D. 152
Schröder, G. 164
science 46–8
securitisation thesis 98
security 40–3, 61–2
Seeger, M.W. 118, 159
Sellnow, T.L. 118
semi-quantitative techniques 21, 26–7, 34, 170, 172
sensemaking 53, 98–9
September 11th 54, 116, 160
sharing of risk 58–9
Shaw, R. 149
Sheffi, Y. 136
Sifaki-Pistolla, D. 140
Simoncini, M. 58
simple learning 152
simulation 86–8
single-loop learning 152–3
situation awareness 52
skills-based learning 151
slow learning 152, 158
Smallman, C. 158
smoking 48
social amplification of risk 26
social construction 25
social impacts 21–2
social learning 26, 161
social media 119–20
societal resilience 134–6
socio-ecological resilience 134–5
solar storms 29–30
Sommer, M. 149, 151–2, 158
Soviet Union 40, 173
space weather 29–30
Spain 164
Sri Lanka 163
Staelraeve, S. 165
stakeholders 14, 19, 23–6, 44, 73, 116–17
standardisation 13, 136–7, 169, *see also* ISO
standby redundancy 83
Stark, A.: learning 152–3, 162, 164; preparedness 70, 74; response 101, 115
statistics 16–17, 29
Stead, E. 158
Sterman, J.D. 157

Stern, E. 72, 86, 101, 115
Stockholm 157
storm damage example 20
Strandberg, J.M. 121
strategic choice situation 106
strategic scenarios 34
stress 110–11, 139–40
subjectivity 97, 162
subsidiarity, principle of 76, 78
supply chain 74, 136
surge capacity 80
Sweden 76, 115, 160–1, 164–5
SWIFT 31–2
Switzerland 158
SWOT analysis 31
systematic mode 25
systemic risks 18–19

't Hart, P. 3, 71, 98, 102, 117, 164–5
tabletop exercises 86
Taiwan 161
technology 30–1, 153; human error and 51–2; prevention 43, 53–4; recovery 137–8; response 114, 120; simulation 87–8
terminating phase 4
Terms of Reference 15
terrorism 54, 80, 116, 157–8, 160, 164
theoretical approaches 4–6, 18, 43, 96, 173–4; to decision-making 103–11; to learning 146, 149, 151–6; to preparedness 72, 82; to risk perception 25–6
Three Mile Island 51, 153
time 3, 21, 33, 72, 101
Tolerable Level of Risk (TLR) 59
tolerance of risk 23–4, 49, 59–60
tort 45, 56
Total Performance Maintenance (TPM) 54
training 43, 53, 86–8, 148
transdisciplinarity 174
triple-loop learning 153
trust, organisational 159
tsunami 32, 50, 120, 163–4
tunnelling, attentional 52

Turner, B.A. 156
Twitter 120
typological-chronological cycle 4–5

UK 82
UK Cabinet Office 86, 128, 149
UN 46, 163
UN International Strategy for Disaster Reduction *see* UNISDR
uncertainty 10–12, 21, 46–7, 62
UNISDR 39, 96, 146; preparedness 69–70, 89; recovery 127–8, 132
Uppsala 103
urban planning 72
US 107, 116, 160, 173; preparedness 74–5, 77, 82, 86; prevention 55, 57; recovery 132, 134–5; response 107, 116
US Department of Homeland Security 132
US National Institute of Technology and Standardization 74–5
USSR 40, 173
Utøya Island 151

Van Asselt, M.B.A. 19, 26, 47
Van de Walle, B. 120
Van Eeten, M. 56–7, 131
Varnado, S. 57
Veil, S.R. 148
Vigsø, O. 121
vocabulary 12–14, *see also* definitions
Vos, E. 47
Vos, F. 140
vulnerability 12, 50–1, 55, 140
Vyncke, P. 117

warning: definitions 88, *see also* early warning
water 24, 53–4
Watters, J. 77, 130–1
Weick, K. 42, 53, 98–9
Westrum, R. 42
Woodard, J. 57

Yoe, C. 24

Zhao, P. 83

Made in the USA
Las Vegas, NV
05 October 2023